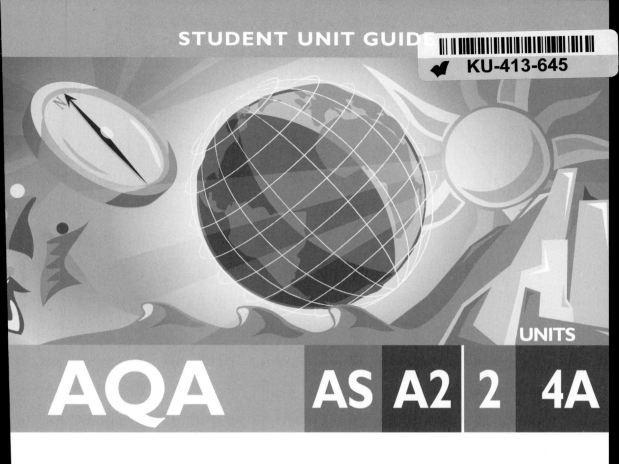

STUDENT UNIT GUIDE

KU-413-645

UNITS

AQA AS A2 2 4A

Geography

Geographical Skills Including Fieldwork

Amanda Barker, David Redfern
and Malcolm Skinner

Philip Allan Updates, an imprint of Hodder Education, an Hachette UK company, Market Place, Deddington, Oxfordshire OX15 0SE

Orders

Bookpoint Ltd, 130 Milton Park, Abingdon, Oxfordshire OX14 4SB

tel: 01235 827827

fax: 01235 400401

e-mail: education@bookpoint.co.uk

Lines are open 9.00 a.m.–5.00 p.m., Monday to Saturday, with a 24-hour message answering service. You can also order through the Philip Allan Updates website: www.philipallan.co.uk

© Philip Allan Updates 2009

ISBN 978-0-340-94799-9

First printed 2009
Impression number 5 4 3
Year 2014 2013 2012 2011

This guide has been written specifically to support students preparing for the AQA AS/A2 Geography Units 2 and 4A examinations. The content has been neither approved nor endorsed by AQA and remains the sole responsibility of the authors.

Printed by MPG Books, Bodmin

Hachette UK's policy is to use papers that are natural, renewable and recyclable products and made from wood grown in sustainable forests. The logging and manufacturing processes are expected to conform to the environmental regulations of the country of origin.

Contents

Introduction

About this guide

All students of A-level geography following the AQA specification have to take Unit 2 Geographical skills at AS. Those who go on to A2 may take Unit 4A Geography fieldwork investigation.

A range of geographical skills and techniques are tested in both these units. In each unit there are also discrete questions on the fieldwork undertaken at AS and A2. This guide provides information to help you with *both* Unit 2 and Unit 4A.

There are three sections to the guide.

Activities

This section contains exemplar activities for most of the geographical skills in the AQA specification for AS and A2. Most of the activities related to AS skills are in the context of either:

- Rivers, floods and their management, or
- Population change

You are advised to try to complete these activities before progressing to the next section.

Answers to activities

Answers are provided for each of the activities in the Activities section.

Fieldwork

This provides:

- information on the fieldwork to be undertaken at AS and A2
- exemplar fieldwork activities for AS and A2
- examples of examination questions on fieldwork and the answers to them, for Unit 2 and Unit 4A, with commentary

Activities
and answers
to activities

This section contains exemplar activities for most of the geographical skills in the AQA specification for both AS and A2. Most of the activities related to AS skills are in the context of either:

- Rivers, floods and their management, or
- Population change

Answers to the activities are provided at the end of this section. You are advised to try to complete the activities before reading the answers.

Activities

Basic skills

Labelling and annotation of photographs

Michael Raw

Figure 1

In the Unit 2 examination you may be asked to label or annotate a photograph such as Figure 1. A **label** is a single word, or short set of words, that simply identifies features shown in the photograph. At this level you are more likely to be asked to annotate a photograph, for example:

(1) Draw an annotated sketch of the river and its valley shown in this photograph.

An **annotation** requires a higher level of labelling which may be detailed description or offer some explanation, or even some commentary.

Annotation of other illustrative material

The same principles apply to other illustrative materials such as base maps, sketch maps, diagrams, graphs and sketches.

(2) Study Figure 2 which shows the location of the highest concentration of post-accession east European migrants in the UK (2006). Annotate the map to show the main features of distribution.

Areas of highest concentration of east European migrants as percentage of population

Figure 2

Use of overlays

Overlays can be physical, such as tracing paper used over photographs, maps or sketches, or electronic, such as those on Windows Live and Google Earth images. It is likely that you will use physical overlays to identify important points on a photograph, map or sketch. The overlay protects the underlying photo, map or sketch from damage. One example of this technique is to identify the main elements of a transport network in an urban area, such as main roads, railways, canals and cycle routes.

Electronic overlays allow you to superimpose (or remove) elements of the human landscape, such as roads, urban areas and other features on (or from) a satellite image of a natural landscape.

You should have experience of both overlay formats by the end of your AS course.

Literacy skills

Comprehension tasks are a regular feature of examination papers. You are required to read such material at speed and digest its main features. An example is given below. Two command words may feature: 'outline' means identify the key features; 'comment on' asks you to infer something from the passage given. You are expected to use your geographical understanding from your studies and apply it to an unfamiliar context. You are being invited to think like a geographer.

(3) Read the information below about the flooding in Morpeth in September 2008. Outline and comment on the economic and social issues that arose.

The streets of Morpeth are buzzing with activity as clearing up continues after the worst flooding in Northumberland for 50 years. Rescue and damage stories can be heard on every corner, while many shops and homes still bear the scars of the weekend's downpour. More than 1,000 properties were affected, and about 400 people were moved to safety when the River Wansbeck burst its banks on Saturday. The area experienced 1 month's rain in just 12 hours. The river returned to its traditional boundaries on Sunday and the town's residents have since been dealing with the aftermath.

The torrent of water was indiscriminate in the victims it chose, with 'closed due to flooding' signs littering the windows of charity shops, restaurants, hairdressers and estate agents. One store owner, unaffected by the flooding, is reporting a boom in sales of mops and brushes. Council officials in fluorescent jackets buzz between the businesses, quickly followed by clip-board-carrying insurance assessors. Carolyn Fisher, 49, owner of Road Runner Sports, is facing up to the possibility of losing 3 months' trade in tough economic times. She said: 'We opened the shop door to a scene of devastation when we returned for the clear-up. All the stock and metal racks were pushed up to the front door and the laminate flooring had cracked and buckled. The more we cleared the more we realised the extent of the damage. The floor was covered in brown sludge and it still really stinks, there must have been sewage in the water.'

Gardens and skips on the residential streets near the River Wansbeck are stacked with mattresses, electrical goods and other household items ruined by water. Evelyn Chapman, 82, Challoner's Gardens, escaped the floods on a raft after being carried out of her home by firefighters. Her bungalow has now been stripped of its waterlogged carpets, furniture and white goods. 'We managed to save my television, DVD and some personal things but it feels like I have lost everything', she said. 'I don't have insurance, so it looks like I will just have to try and save some money up to buy more things.' As rain begins to fall again on Tuesday thoughts naturally turn to the prospect of further flooding. However, Mrs Chapman's 49-year-old daughter Eileen responds to the idea of another flood in good humour. 'Why worry, she has nothing more to lose', she quips.

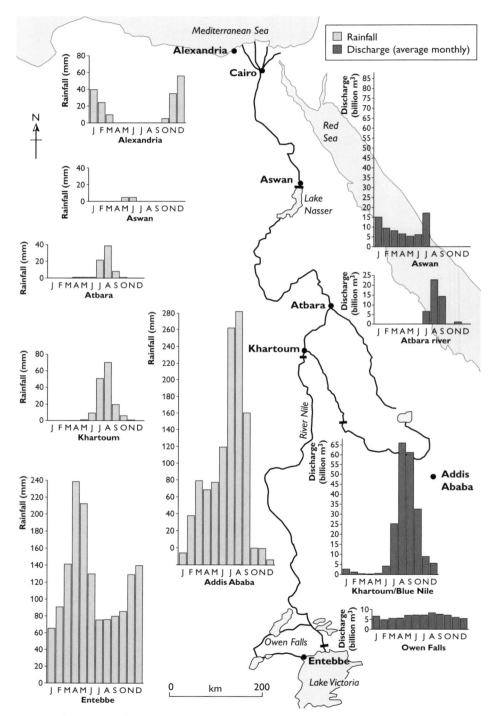

Figure 3 Rainfall and discharge at selected sites in the Nile basin

Cartographic skills

Maps with located proportional symbols

(4) (a) **Complete Figure 3 using the discharge data for Aswan in Table 1.**

(b) **What is the relationship between the discharge at Aswan and the rainfall that occurs within the Nile basin?**

Table 1

Month	Monthly flow (billion m^3)
August	65
September	75
October	45
November	28
December	19

Maps with located proportional circles

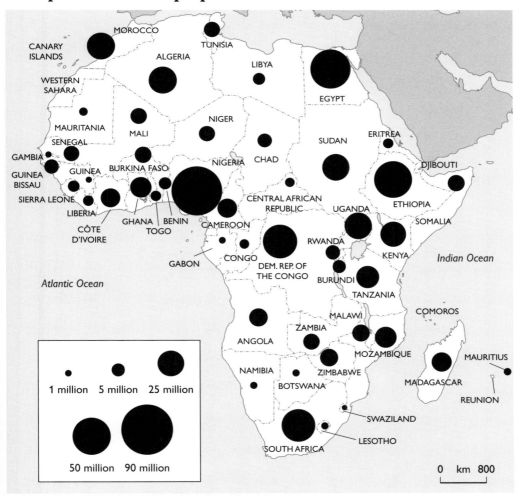

Figure 4 Population distribution in Africa, 1990

(5) (a) **Give the approximate population in 1990 of:**
    (i) Nigeria   **(iii) Sierra Leone**
    (ii) Egypt   **(iv) Ethiopia**

   (b) **Why do you need to exercise great care when constructing maps of this type?**

Flow lines

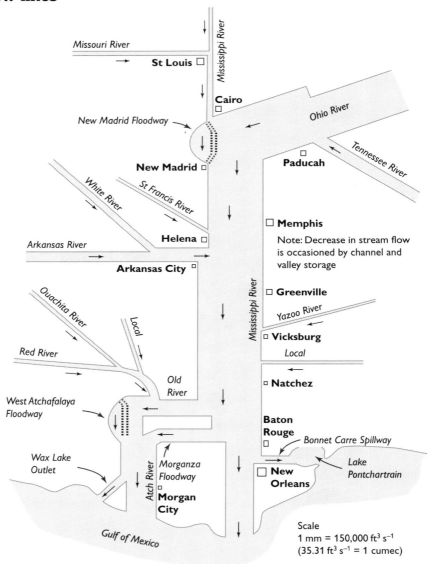

Figure 5 Project Design Flood, 1958

(6) After disastrous floods in the 1920s and 1930s in the lower Mississippi Basin, the US Corps of Engineers was given the task of drawing up a flood prevention scheme. **Figure 5** shows the forecast of the maximum flow through the scheme which the Corps of Engineers had designed to control flooding on the Mississippi.

(a) What was the approximate projected maximum flow (discharge) of:
    (i)   the River Ohio after its confluence with the Tennessee?
    (ii) the River Arkansas above its confluence with the White River?
    (iii) the Mississippi at Vicksburg?
    (iv) the Mississippi at New Orleans?

(b) What are the advantages of this method (flow diagrams) for displaying data on river discharge?

Choropleth maps

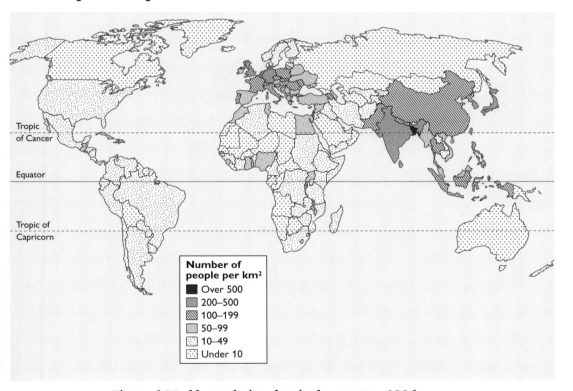

Figure 6 World population density by country, 1996

(7) (a) Describe the steps you would take in order to construct a map such as Figure 6.

(b) What are the advantages and disadvantages of presenting data in this form?

Isolines

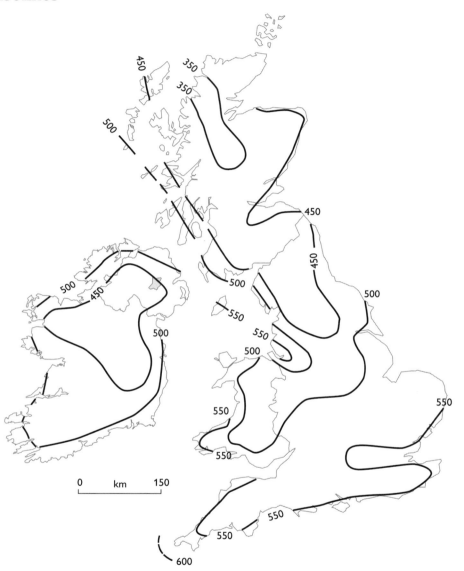

Figure 7 Potential evapotranspiration in the British Isles (mm)

(8) (a) **Complete Figure 7 by drawing the 400 mm line.**

   (b) (i) **What do you understand by the term 'potential evapotranspiration'?**

   (ii) **Describe the pattern of evapotranspiration over the British Isles.**

Graphical skills

Line graphs/bar graphs

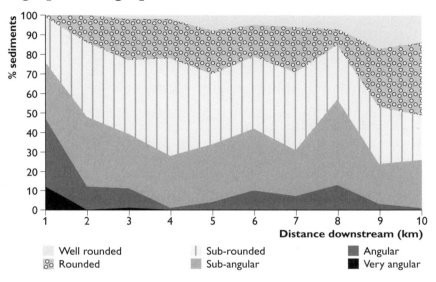

% sediments (y-axis: 0, 10, 20, 30, 40, 50, 60, 70, 80, 90, 100)

Distance downstream (km) (x-axis: 1 to 10)

Well rounded   Sub-rounded   Angular
Rounded   Sub-angular   Very angular

Figure 8 Downstream changes in sediment shape on the River Aire
(Malham to Gargrave)

(9) (a) From Figure 8, calculate the percentage of different types of sediment (i.e. well rounded, rounded, sub-rounded, sub-angular etc.) at the following points on the River Aire:

    (i) 1 km downstream

    (ii) 6 km downstream

(b) (i) Describe the changes in sediment shape that occur between **Malham and Gargrave**, along the **River Aire**.

    (ii) Attempt to explain the changes which occur between 7 and 9 km along the section.

(10) (a) Complete Figure 9 by drawing the graph for the discharge of the River Lagan from the figures in Table 2.

(b) Compare the responses of the three rivers to the rainfall which fell from 15 to 22 October 1987.

Table 2

15 October	4 cumecs
16 October	4 cumecs
17 October	9 cumecs
18 October	17 cumecs
19 October	16 cumecs
20 October	10 cumecs
21 October	85 cumecs
22 October	123 cumecs

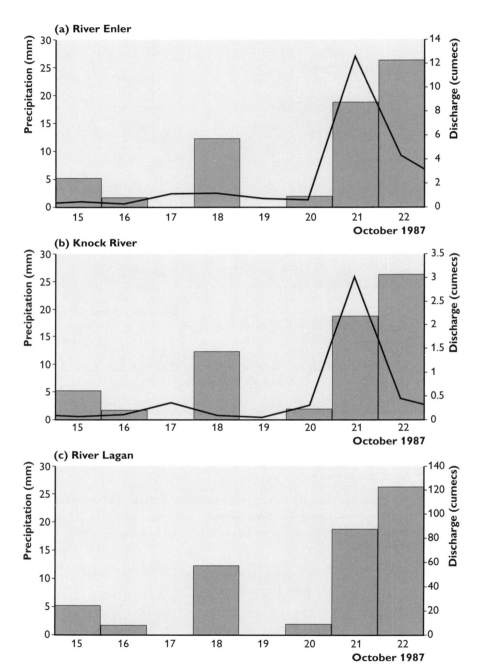

Figure 9 Flood discharges (cumecs) of three rivers in Northern Ireland, all responding to the same rainfall in October 1987. (a) River Enler, (b) Knock River, (c) River Lagan

Scattergraphs

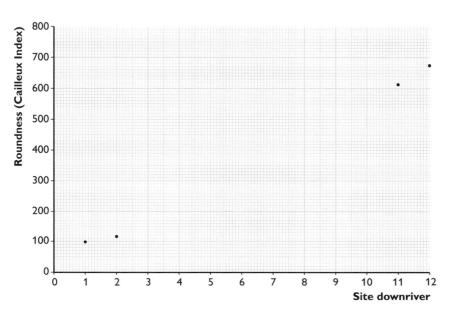

Figure 10 Pebble roundness on the River Kent (Cumbria)

Table 3

(11) (a) (i) During a survey of the River Kent (Cumbria), measurements were made on the bedload to establish its angularity/roundness. The Cailleux Index was calculated, in which the higher the figure the more rounded the pebble.

The observations given in Table 3 were made.

Draw a scattergraph to show the relationship between pebble roundness and distance down the river. Sites 1, 2, 11 and 12 have been plotted for you on Figure 10.

Site	Cailleux Index
Site 1	98.00
Site 2	113.00
Site 3	136.00
Site 4	158.00
Site 5	203.13
Site 6	285.11
Site 7	330.49
Site 8	424.36
Site 9	467.14
Site 10	550.00
Site 11	615.52
Site 12	674.07

(ii) Insert a best-fit line on the completed graph.

(b) Describe the relationship between these two variables and attempt to explain the relationship that you have described.

Proportional divided circles (pie charts)

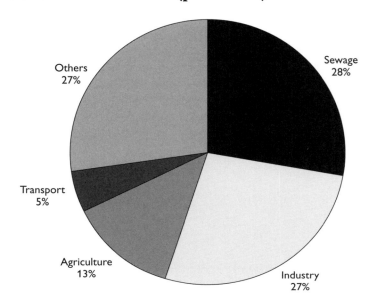

Note: 'Others' are incidents in which the
source cannot be accurately identified.

Figure 11 River pollution incidents in England and Wales, 1990

(12) (a) **Draw a pie chart to show the sources of pollution incidents in English and Welsh rivers in 1996. The figures are given in Table 4.**

Table 4

Pollutant	Per cent of total
Sewage	30
Industry	20
Agriculture	12
Transport	8
Others	30

(b) **Describe the changes in the source of river pollution in England and Wales that occurred between 1990 and 1996.**

activities

Triangular graphs

① Carl Beck, North Pennines

② Great Eggleshape Beck, mid-Wales

③ Upper Wye, Welsh Borders

④ Preston Montfort

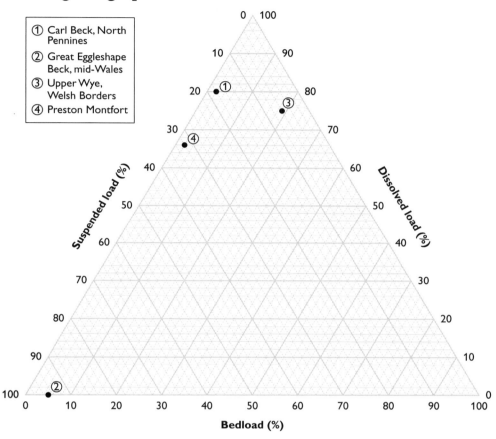

Figure 12 The sediment load of selected British rivers

(13) (a) (i) Calculate the share of the total load made up from dissolved load, suspended load and bedload for all four rivers shown in Figure 12.

(ii) Complete the graph by adding the data for the River Farlow (Shropshire) given in Table 5.

Table 5

Dissolved	50%
Suspended	46%
Bedload	4%

(b) What are the advantages of using triangular graphs to represent data?

Kite diagrams

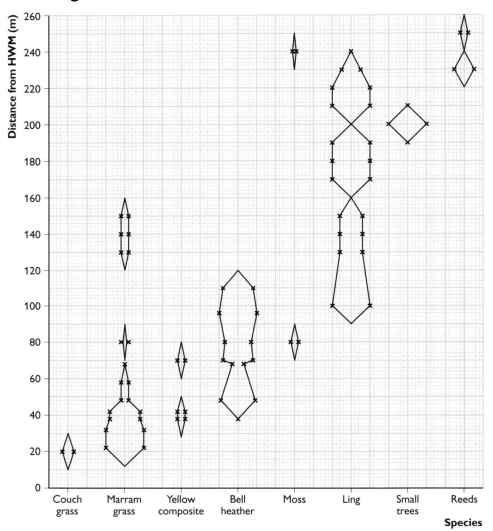

Figure 13 Vegetation changes along a dune transect in Studland Bay, Dorset

(14) (a) Describe:
  (i) the distribution of marram grass across the dune transect
  (ii) the vegetation that occurs at 80 m across the transect

  (b) What are the advantages of displaying vegetation transect data on this type of diagram?

Graphs using logarithmic scales

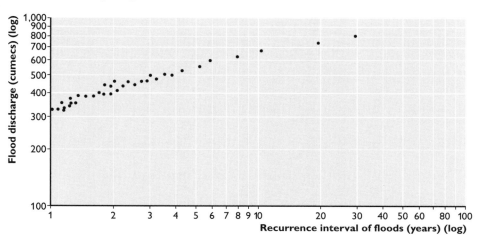

Figure 14 Flood frequency of the River Wye at Hereford, 1908–1999

(15) (a) Complete Figure 14 by plotting the flood of 4 December 1960 on the River Wye when the discharge was 960 cumecs. This was estimated to be a once in every 60 years event.

   (b) (i)   What would be the magnitude of the 10-year flood and the 50-year flood?

   (ii) Describe the relationship between flood discharge and recurrence interval.

Dispersion diagrams

See 'Statistical skills' below.

Statistical skills

Measurement of central tendency: mean, median and mode

(16) Students studying the differences between an upstream and a downstream site on the River Eea in the Lake District, Cumbria measured the long axis of 15 pebbles from the bed of the river in each location.

Table 6

Upstream (cm)	Downstream (cm)
15.0	4.0
8.0	8.0
22.0	10.0
32.0	6.0
16.0	19.0
18.5	14.0
34.0	6.0
32.0	13.5
19.5	7.0
13.5	5.0
28.0	12.5
10.5	12.0
13.0	8.5
24.5	6.0
45.0	13.0

(a) **Calculate the mean, median and mode long-axis value for the upstream and downstream sites.**

(b) **Why do the mean values differ between the upstream and downstream sites?**

(c) **Which of these indices is most useful to students studying the difference between the characteristics of bedload between an upstream and a downstream site?**

Measurement of dispersion: dispersion diagrams, inter-quartile range and standard deviation

The data in Table 6 show considerable variation in the sizes of the pebbles measured at the upstream and downstream sites of the River Eea. It was shown in Question 16 that the median and mean values for each site were different, as is usual for data collected in the field. The larger the difference between the two, the more a set of data is said to be skewed.

There are a number of techniques that can be used to study the range of values in a set of observations. These include dispersion diagrams, calculation of the inter-quartile range and standard deviation.

(17) (a) **Complete a dispersion diagram for the upstream and downstream sites on the River Eea, using the data from Table 6. This will show the range of pebble sizes for the downstream and upstream sites.**

(b) **The inter-quartile range is a measure of the spread of values around their median. The greater the spread, the higher the inter-quartile range. Calculate the inter-quartile range for each set of data.**

- Start by using the dispersion diagram showing the long-axis values ranked in order of size for the upstream site, putting the smallest first and the largest last.
- Identify the median value.
- Next, work out the upper quartile value by identifying the mid-point (upper median) from the seven values in the upper half of the dispersion diagram.
- Now work out the lower quartile by repeating the above but with the lower half of the dispersion diagram.
- The inter-quartile range can now be calculated by taking the lower quartile value away from the higher quartile value.
- Repeat this for the downstream site.

(c) How do the inter-quartile ranges differ between the upstream and downstream sites on the River Eea?

(d) In a normal distribution most values are close to the mean. Standard deviation (σ) is another method that can be used to help measure the dispersal of the observations around the mean. If σ is small then the pebbles show little variation in size, if σ is relatively large then the pebbles show great variation. Use Table 7 to help calculate the standard deviation for the upstream and downstream sites.

$$\sigma = \sqrt{\frac{\Sigma(\bar{x} - x)^2}{n}}$$

Table 7

Upstream (cm)		Downstream (cm)	
x	x^2	y	y^2
15		4	
8		8	
22		10	
32		6	
16		19	
18.5		14	
34		6	
32		13.5	
19.5		7	
13.5		5	
28		12.5	
10.5		12	
13		8.5	
24.5		6	
45		13	
$n = 15$	$\Sigma x^2 =$	$n = 15$	$\Sigma y^2 =$

(e) Compare the calculated standard deviation values for the upstream site (x) and the downstream site (y). What do these values tell you about the bed-load samples collected?

Spearman rank correlation coefficient and the application of significance testing

(18) Students measured a range of variables along the River Eea at ten sites between its source and mouth. They wanted to test whether the Bradshaw model could be applied to this river. The Spearman rank correlation coefficient test can be used to help determine whether or not the River Eea conforms to the Bradshaw model. (To use Spearman rank the data must be ordinal, not put into categories.)

$$R_s = 1 - \left(\frac{6 \, \Sigma d^2}{n^3 - n} \right)$$

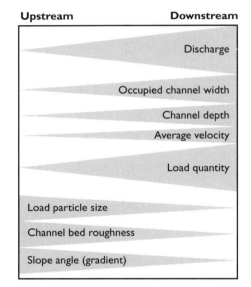

Figure 15 The Bradshaw model

Table 8

Site no.	Width (m)	Rank	Average depth (m)	Rank	d	d²
1	0.97		8.14			
2	1.30		6.00			
3	2.01		8.12			
4	3.06		17.10			
5	3.50		16.50			
6	2.55		14.68			
7	6.23		8.30			
8	6.00		9.40			
9	6.42		24.70			
10	7.25		20.91			
						$\Sigma d^2 =$

n = number of data pairs, in this case n = 10
d = difference between the ranks of each of the paired variables
Σ = sum of the values, in this case the total of all the d^2 values added together

According to the Bradshaw model there should be a relationship between channel width and depth.

(a) State the expected and null hypotheses that might be tested for these channel variables.

(b) Use Table 8 to calculate the Spearman rank correlation coefficient. Start by ranking the width values from highest (1) to smallest (10). Then rank the average depth column in the same way.

(c) Once the value of R_s has been calculated, the result must be tested for significance, because there is a possibility that the result could have occurred by chance alone. (Significance is hard to prove if there are fewer than eight pairs of data, as the sample size is too small.)

- Decide on the rejection level. This tells you how certain you want to be that the result did not occur by chance. This is given in a table and is usually 0.05 or 95% sure, or 0.01 or 99% sure.
- Look up the critical value in Table 9. Degrees of freedom is the number (n) of paired measurements in your sample.

Table 9

n	Rejection level		
	0.05	0.02	0.01
5	1	1	
6	0.886	0.943	1
7	0.786	0.893	0.929
8	0.738	0.833	0.881
9	0.683	0.783	0.833
10	0.648	0.746	0.794
12	0.591	0.712	0.777

(d) Now that the R_s value has been calculated and tested for significance decide if you will accept or reject the null hypothesis.

(e) To what extent does this result help you to decide whether or not the River Eea conforms to the Bradshaw model?

Comparative tests

Chi-squared test

$$\chi^2 = \sum \frac{(O - E)^2}{E}$$

(19) With this test it is possible to look for an association between two sets of data which have been put into categories. When studying the River Eea students measured the roundness of pebbles at Site 2 (upstream) and at Site 8 (downstream) using Powers' scale of roundness. Their results are shown in Table 10.

Null hypothesis: There is no difference in the shape of pebbles between an upstream and downstream location on the River Eea. They are distributed randomly.

Table 10

Row number (R)	Column number (K)	Upstream O	Upstream E	Downstream O	Downstream E	ΣR
		K1		K2		
R1	Angular	22		7		29
R2	Sub-angular	15		2		17
R3	Sub-round	9		10		19
R4	Round	4		31		35
		$\Sigma K1 = 50$		$\Sigma K2 = 50$		$n = 100$

R = row number.
K = column number.
O = the observed frequency of pebbles in each category.

(a) **Calculate the expected frequency (E) in each cell using the formula:**

$$E = \left(\frac{\Sigma R \ \Sigma K}{n} \right)$$

(b) **Calculate the chi-squared value using the formula.**

(c) **Now check this result with the significance table in a book of statistical tables. First calculate the degrees of freedom, $(R - 1) \times (K - 1)$. Then read from the critical values table (Table 11) to see if the null hypothesis can be rejected. Your chi-squared value must be greater than the critical value to be significant.**

Table 11

Critical value: 0.05		Critical value: 0.01	
Degrees of freedom	Value	Degrees of freedom	Value
1	3.84	1	6.63
2	5.99	2	9.21
3	7.82	3	11.30
4	9.49	4	13.30
5	11.10	5	15.10

(d) **What does the result tell us about the two samples of pebbles taken from the River Eea?**

Mann–Whitney U test

This test can be used to make a comparison of two medians. It also determines whether this difference is significant. Unlike the chi-squared test, actual values not categories are used. Both sample sizes should be below 20 but above 5. Each sample does not have to have the same number of observations.

(20) The long axis of pebbles for 15 samples was measured at the upstream and downstream sites on the River Eea in Cumbria.

Null hypothesis: There is no significant difference between pebble sizes at the upstream and downstream sites on the River Eea.

$$U_x = N_x N_y + \frac{N_x(N_x + 1)}{2} - \Sigma r_x$$

$$U_y = N_x N_y + \frac{N_y(N_y + 1)}{2} - \Sigma r_y$$

N_x and N_y are the two sample sizes and Σr_x is the sum of the ranks for sample x.

(a) Use Table 12 to set out the ranks in order (lowest value first). Remember that the total sample is ranked together and not as individual columns as in the Spearman rank correlation coefficient.

Table 12

Upstream		Downstream	
x	Rank (r_x)	y	Rank (r_y)
15		4	
8		8	
22		10	
32		6	
16		19	
18.5		14	
34		6	
32		13.5	
19.5		7	
13.5		5	
28		12.5	
10.5		12	
13		8.5	
24.5		6	
45		13	
No. in sample (N_x) = 15		No. in sample (N_y) = 15	
Total of rank scores (Σr_x) =		Total of rank scores (Σr_y) =	

(b) Calculate the U values for both the upstream site (x) and the downstream site (y) using the formula.

(c) In order to test the result you must now test for significance. Take the smaller U value calculated and consult Table 21 (p. 44) to decide whether the null hypothesis can be rejected.

(d) Explain the result.

Answers to activities

Basic skills

Labelling and annotation of photographs

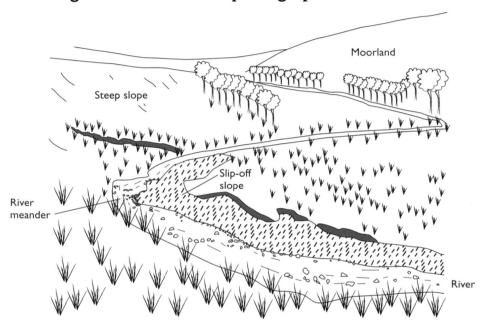

Figure 16

(1) Annotation requires a higher level of labelling, which may be detailed description, some explanation, or even some commentary. For example, in relation to the sketch in Figure 16, annotations could include:
- a gentle slip-off slope consisting of larger deposits such as cobbles and pebbles (description)
- an incised meander caused by rejuvenation of the river (explanation)
- an area of heather moorland which may be used for grouse shooting (commentary)

Annotation of other illustrative material

(2) Annotations for Figure 2 in the question could include:
- very high proportions of eastern European migrants in the north of Scotland, though the actual numbers involved are likely to be low
- concentrations in Fenland areas, and Hereford and Worcester — due to agricultural labour needs?

- some evidence of concentration in coastal tourist areas — employment in hotels, and/or care homes in these areas
- concentrations in industrial areas of the north, west midlands and the London area

Literacy skills

(3) An *outline* of the economic and social issues raised in the Morpeth extract may include:

- **Economic:** damage to houses, shops and other services; loss of earnings from the closure of businesses; high payments from insurance; loss of savings of those who were uninsured.
- **Social:** people having to evacuate their homes and shops; people may be made redundant because of loss of trade; dirt, sewage and smells will need to be cleaned up/removed; loss of household goods and personal items.

Commentary could include:
- it could be a long time before the area returns to any sense of normality
- people may need to be re-housed for a prolonged period of time
- in the immediate aftermath, there could have been a threat of disease from the sewage
- insurance premiums will increase and some people may not be able to get insurance
- the local council and the Environment Agency will need to assess future flood risk and protection methods

Cartographic skills

Maps with located proportional symbols

(4) (a) See Figure 17 on next page.

(b) The discharge at Aswan is low through the first 6 months with decreasing monthly totals, but increases dramatically through August, September, October, November and December. Peak discharge is in September. The high rainfall in the southern part of the basin (figures for Entebbe) isn't reflected in the discharge because as rainfall increases in the south, the discharge at Aswan is actually decreasing. This rainfall, however, will take some time to pass down the river, and the large lakes of the south, particularly Lake Victoria, act as natural reservoirs releasing water at an even rate.

The other tributaries of the Nile rise in areas where there is substantial rainfall from May onwards, and this is reflected in the flood season at Aswan. The peak rainfall in those regions is in July and August, giving a peak discharge at Aswan in August, September and October. As rainfall decreases in those regions, so the river gradually falls to a much lower level. The rainfall in the Aswan region is so sparse that it can have little or no influence on the discharge of the Nile.

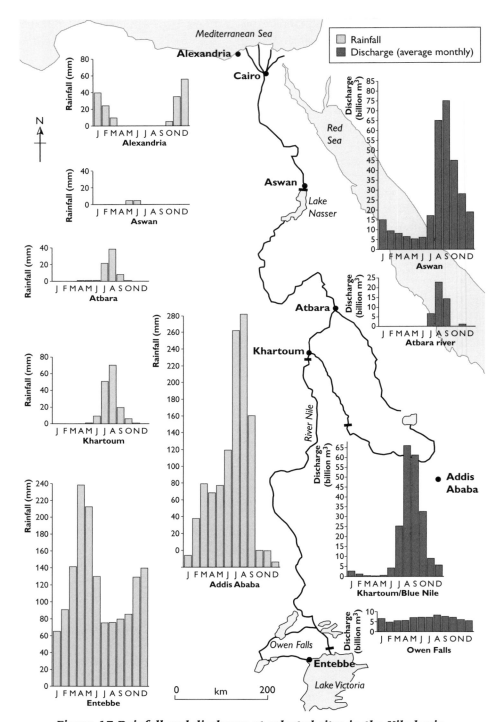

Figure 17 Rainfall and discharge at selected sites in the Nile basin

Maps with located proportional circles

(5) (a) (i) 90 million

      (ii) 50 million

      (iii) 5 million

      (iv) 50 million

(b) Because of the varying size of each plot it is often difficult to foresee the end product. Careful planning and care in the placing of symbols is therefore needed to avoid overlap and confusion. It is essential that each symbol has 'a sense of place' in that it must be clear which area the symbol represents.

Flow lines

(6) (a) (i) 2,250,000 ft^3 s^{-1} (2,150,000–2,350,000)

      (ii) 400,000 ft^3 s^{-1} (350,000–450,000)

      (iii) 2,550,000 ft^3 s^{-1} (2,450,000–2,650,000)

      (iv) 1,050,000 ft^3 s^{-1} (1,175,000–1,325,000)

(b) The obvious advantage is that the flow of water is represented by a flow diagram, which gives a straightforward visual impression of the actual flow of the river. It is also possible to see the sequence, i.e. how flows from various rivers come together to form the larger main river. In the case of the Mississippi, it is also possible to see the amount of water that is diverted away from the main channel as one of the elements of flood control.

    As the width of each line is proportional to the actual movement of water, flows can be measured at any point of the sequence and relationships between the various elements which make up the whole can be seen and also measured. Although the diagram has many of the attributes of a map, it is actually topological: elements are in the correct sequence, but real distances (and direction) are distorted in order to achieve a clearer image.

Choropleth maps

(7) (a) Step 1 Decide on the number of classes needed to display the full range of data.

    Step 2 Select suitable class boundaries and the range of each class (the class interval).

    Step 3 Devise a key for the density of shading appropriate to each class. The highest density should have the darkest shading.

    Step 4 Do not represent the lowest density by no shading, i.e. blank. This gives the impression that the figure for the area is zero. Use 'blank' for areas where there are no data available.

    Step 5 Shade in each area of the map with the appropriate form of shading from the key.

(b) The main advantages are that the maps are easy to construct and are visually effective as they enable the reader to see general patterns in an area distribution. The limitations of the method are that the maps assume the whole area has the same value, but there could be important variations within an area. In Figure 6, for example, Brazil has one level of shading indicating a low density (10–49 people per km²), whereas in reality the coastal fringes of the country have a much higher population density than the interior. Choropleth maps also show abrupt changes in values at boundaries of areas, which in most cases are not accurate.

Isolines

(8) (a)

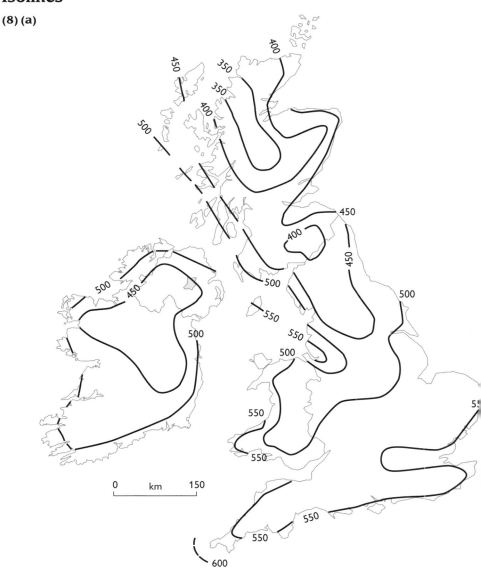

Figure 18 Potential evapotranspiration in the British Isles (mm)

(b) (i) The combined effect of evaporation (from surfaces) and transpiration (from vegetation) is known as evapotranspiration. Potential evapotranspiration is the water loss that would occur if there was an unlimited supply of water in the soil.

(ii) Potential evapotranspiration (PEVT) is highest in the south of the British Isles and lowest in the north. The isolines do not run east to west, though, as there are differences between coastal areas and those inland. PEVT tends to be higher on the same latitude in those areas that are on the coast — coastal northern England has a level of 500–550 mm, whereas corresponding areas inland have 400–450 mm. Coastal Ireland is around 500 mm, and its inland areas have 450 mm or less. There is a large area below 350 mm in northern Scotland, running from the interior to the northern coast. The highest area in the British Isles is off the southwest of England at 600 mm.

Graphical skills

Line graphs/bar graphs

(9) (a) (i)

Very angular: 11%	Sub-angular: 30%	Rounded: 2%
Angular: 34%	Sub-rounded: 23%	Well-rounded: 0%

(ii)

Very angular: 0%	Sub-angular: 33%	Rounded: 16%
Angular: 10%	Sub-rounded: 37%	Well-rounded: 4%

(b) (i) At the beginning of the section near Malham the sediment is dominated by angular rather than rounded sediments. Angular pieces account for 75% of all sediments and there are no well-rounded particles in the river. Very angular pieces are only seen in the first kilometre and angular fragments in general decline rapidly, so much so that by 3 km they are no longer in the majority. Well-rounded pieces make an appearance between 2 and 3 km and rounded fragments as a whole continue to dominate until beyond 7 km, where angular pieces are in the majority for a short distance. This situation does not last long and the rounded sediment again continues to dominate beyond that point. At Gargrave, at the end of the section, rounded pieces form 70% of all sediment, a complete reversal from the situation at the beginning near Malham.

(ii) As there is an increase in the angularity of the sediment at this point, it is clear that angular fragments must have been added to the River Aire. One possible explanation of a source for such sediment is a tributary river which enters at this point carrying a higher level of angular fragments than the main river. This sediment is gradually eroded by attrition as it passes down the main river, so the rounded fragments are again in the majority downstream of this point. (Note that in fact this is the point where Otterburn Beck joins the River Aire.)

(10) (a)

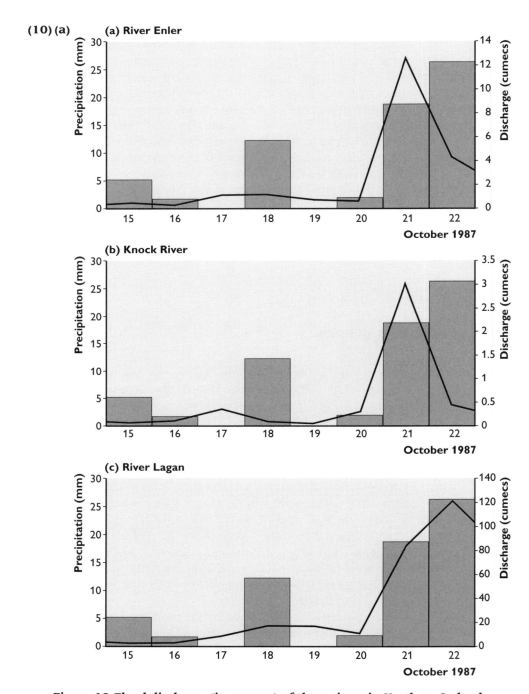

*Figure 19 Flood discharge (in cumecs) of three rivers in Northern Ireland,
all responding to the same rainfall in October 1987. (a) River Enler,
(b) Knock River, (c) River Lagan*

(b) All three rivers respond to the rainfall of 15 and 16 October by rising slightly late on 16 October and through into 17 October. The Enler and Lagan maintain this flow into 18 October whereas the Knock has a sharp fall back to its previous levels. The Enler continues to fall through to 20 October, as does the Lagan, but the Knock clearly responds on 20 October to that day's rainfall. The rainfall of 21 and 22 October is large (over 40 mm) and as the catchments are obviously saturated from the previous rainfall, there is an immediate response from all three rivers. The Enler and Knock peaked on 21 October, but the Lagan had a slower response, peaking on the next day, 22. All three rivers have a flashy regime, with the Knock and Enler very quick to respond to rainfall events.

Scattergraphs

(11)(a) (i)
(ii)

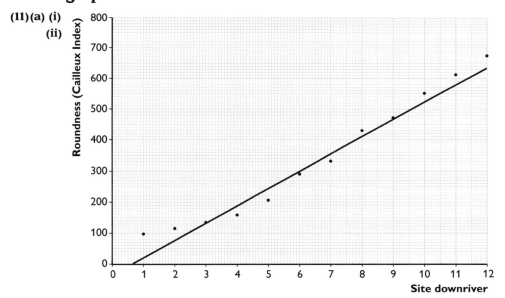

Figure 20 Pebble roundness on the River Kent (Cumbria)

(b) The graph shows that the further downstream you go on the River Kent, the more rounded the bedload will be.

Particles eroded from the channel, together with materials input by mass movements and weathering from valley slopes, are transported downstream. Sediment in transport is used by the river in abrasion to wear away the bed and sides of the channel. Pebbles are therefore eroded by this action and also by colliding with each other. This process, known as attrition, not only reduces pebbles in size, but also makes them smoother by removing hard edges. The further a river runs, therefore, the more rounded its load becomes.

Proportional divided circles (pie charts)

(12)(a)

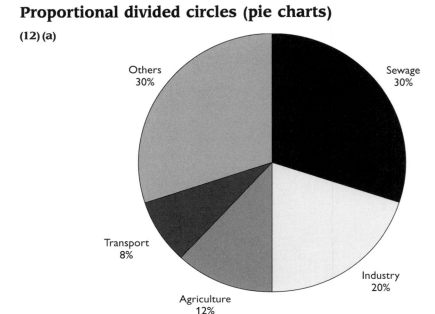

Figure 21 River pollution incidents in England and Wales, 1996

(b) Overall, the changes between 1990 and 1996 have been only small. Sewage incidents appear to have increased slightly. Incidents caused by industry have declined, in percentage terms more than the increase in sewage. Agricultural and transport incidents show little change, although transport has demonstrated an observable increase. 'Others' has slightly increased. The problem with this information is that around one-third of all incidents are in a category for which the source cannot accurately be defined, which is difficult for making overall assessments.

Triangular graphs

(13)(a)(i) The percentages are shown in Table 13.

Table 13

River	Dissolved	Suspended	Bedload
Carl Beck	80	18	2
Great Eggleshape	0	95	5
Upper Wye	75	6	19
Preston Montfort	66	32	2

(ii)

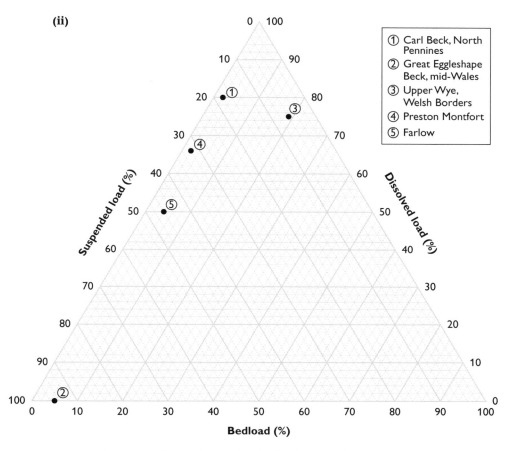

Figure 22 *The sediment load of selected British rivers*

(b) The advantages of using triangular graphs are:
- very useful for displaying data when the whole can be broken down into three components (such as the river sediment data in this question)
- the varying proportions can be seen, indicating the relative importance of each
- the dominant component in each case can be seen
- after plotting, clusters will sometimes emerge, enabling a classification of the items involved

Kite diagrams

(14)(a)(i) Marram grass appears early in the transect. Along with couch grass, it is the first type of vegetation to colonise the area, being seen at around 12 m into the transect. Marram dominates the dunes from the 12 m mark up to 40–50 m where it begins to decline. It is overshadowed by other plants from then on. Nevertheless it survives well into the transect, a second area of the vegetation being found between 120 and 160 m, although its extent is much reduced from that of its location closest to the sea.

(ii) There are three types of vegetation present at a point 80 m from the sea. The area is dominated by bell heather, which covers an area over three times greater than any other plant. The second most important plant is moss, which covers twice the area of the final plant, marram grass. Some yellow composite was found just before the 80 m mark, but it was not present on the mark itself.

(b) Kite diagrams are useful for displaying changes over distance as one axis can be used for distance and the other for individual plant species. The width of the kite, representing a single species, enables a visual comparison to be made of the distribution of vegetation at any point in the section.

Graphs using logarithmic scales

(15)(a)

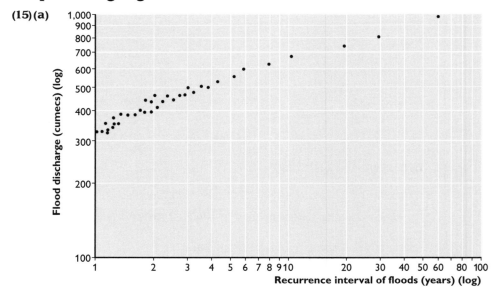

Figure 23 Flood frequency of the River Wye at Hereford, 1908–1990

(b) (i) The magnitude of a 10-year flood would be over 600 cumecs.
The magnitude of a 50-year flood would be over 850 cumecs.

(ii) There is a positive correlation between the flood magnitude (measured by river discharge) and the number of years between such events (recurrence interval). This graph can then be used to predict the likely recurrence of a flood of a particular size on the River Wye.

Dispersion diagrams

See 'Statistical skills'.

Statistical skills

Measurement of central tendency: mean, median and mode

(16)(a) Upstream: the mean value for the long axes of bedload sampled is 21.1 cm, the median value is 19.5 cm and the mode 32 cm.

Downstream: the mean value is 9.63 cm, the median 8.5 cm and the mode 6.0 cm.

(b) The mean value for the upstream site demonstrates that pebbles are generally larger. This is because in an upland channel, bedload originates from the valley sides as the products of recent weathering and mass movement. An upland stream has a low discharge for most of the time, consequently its competence is relatively low and the larger material which has rolled into the channel can only be moved downstream during incidences of flash flooding. Smaller bedload is evident in a downstream channel for two reasons. First, material has been transported downstream by processes such as traction and saltation. Second, this material will have been worn down by the erosion process of attrition.

(c) The mean, median and mode are all useful, although crude, tools when comparing sets of data. However, the mean allows for further mathematical processing using **standard deviation**, and median values are also considered further when calculating the **inter-quartile range**. The median and mode are quick to calculate and are unaffected by extreme values; the mean can fall between a very wide range of data values.

Measures of dispersion: dispersion diagrams, inter-quartile range and standard deviation

(17)(a)

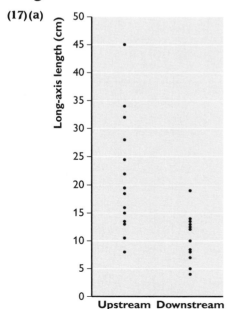

Figure 24 Dispersion diagram showing pebble sizes upstream and downstream on the River Eea

(b) *Table 14*

Upstream (cm)		Downstream (cm)	
45.0		19.0	
34.0		14.0	
32.0		13.5	
32.0	**Upper quartile**	**13.0**	**Upper quartile**
28.0		12.5	
24.5		12.0	
22.0		10.0	
19.5	**Median**	**8.5**	**Median**
18.5		8.0	
16.0		7.0	
15.0		6.0	
13.5	**Lower quartile**	**6.0**	**Lower quartile**
13.0		6.0	
10.5		5.0	
8.0		4.0	

(c) Upstream IQR 32.0 – 13.5 = 18.5 cm. Downstream IQR 13.0 – 6.0 = 7.0 cm

(d) Standard deviation

Table 15

Upstream (cm)		Downstream (cm)	
x	x^2	y	y^2
15	225	4	16
8	64	8	64
22	484	10	100
32	1,024	6	36
16	256	19	361
18.5	342.25	14	196
34	1,156	6	36
32	1,024	13.5	182.25
19.5	380.25	7	49
13.5	182.25	5	25
28	784	12.5	156.25
10.5	110.25	12	144
13	169	8.5	72.25
24.5	600.25	6	36
45	2,025	13	169
$n = 15$	$\Sigma x^2 = 8{,}826.25$	$n = 15$	$\Sigma y^2 = 1{,}642.75$
	mean $x = 22.1$		mean $y = 9.63$
$\sigma = \sqrt{\dfrac{8{,}826.25}{15} - 488.41}$		$\sigma = \sqrt{\dfrac{1{,}642.75}{15} - 92.74}$	
$\sigma = 10.00$		$\sigma = 4.09$	

(e) Upstream site: This result means that there is a 68% probability of the pebble's long axis lying between 12.1 cm and 32.1 cm. Downstream site: This result means that there is 68% probability of the pebble's long axis lying between 5.54 cm and 13.72 cm.

Spearman rank correlation coefficient and the application of significance testing

(18)(a) Expected hypothesis: There is a relationship between channel width and depth. Null hypothesis: There is no relationship between channel width and channel depth.

(b)

Table 16

Site no.	Width (m)	Rank	Average depth (m)	Rank	d	d^2
1	0.97	10	8.14	8	2	4
2	1.30	9	6.00	10	−1	1
3	2.01	8	8.12	9	−1	1
4	3.06	6	17.10	3	3	9
5	3.50	5	16.50	4	1	1
6	2.55	7	14.68	5	2	4
7	6.23	3	8.30	7	−4	16
8	6.00	4	9.40	6	−2	4
9	6.42	2	24.70	1	1	1
10	7.25	1	20.91	2	−1	1
						$\Sigma d^2 = 42$
						$6\Sigma d^2 = 252$

$$R_s = 1 - \left(\frac{6 \, \Sigma d^2}{n^3 - n} \right) = 1 - \left(\frac{252}{990} \right) = 1 - 0.2545$$

$R_s = 0.7455$, a positive relationship

(c) The degrees of freedom in this calculation are 10 (10 pairs of data).

Table 17

	Rejection level		
n	0.05	0.02	0.01
5	1	1	
6	0.886	0.943	1
7	0.786	0.893	0.929
8	0.738	0.833	0.881
9	0.683	0.783	0.833
10	0.648	0.746	0.794
12	0.591	0.712	0.777

(d) Using the significance table we can see that the value of R_s (0.7455) is greater than the critical values at the 0.05 level. As a result we can be more than 95% sure that the value calculated did not occur by chance, so we can reject the null hypothesis.

As the calculated value is lower than the critical value of 0.794 at the 0.01 or 99% level of significance, we cannot reject the null hypothesis at this higher level.

To summarise, the data show that there is a positive correlation between average width and depth, and there is 95% certainty that this result did not occur by chance.

(e) The Bradshaw model suggests that there is a positive correlation between width and depth, and that both increase with distance downstream. The Spearman rank correlation coefficient proves that the River Eea in Cumbria conforms to the model because a statistically significant positive relationship was established between these two sets of variables.

Comparative tests

Chi-squared test

(19) (a) *Table 18*

Row number (R)	Column number (K)	Upstream O	Upstream E	Downstream O	Downstream E	ΣR
		K1		K2		ΣR
R1	Angular	22	14.5	7	14.5	29
R2	Sub-angular	15	8.5	2	8.5	17
R3	Sub-round	9	9.5	10	9.5	19
R4	Round	4	17.5	31	17.5	35
		ΣK1 = 50		ΣK2 = 50		n = 100

(b) $\chi^2 = \dfrac{(22 - 14.5)^2}{14.5} + \dfrac{(7 - 14.5)^2}{14.5} + \dfrac{(15 - 8.5)^2}{8.5} + \dfrac{(2 - 8.5)^2}{8.5} + \dfrac{(9 - 9.5)^2}{9.5} + \dfrac{(10 - 9.5)^2}{9.5} +$

$\dfrac{(4 - 17.5)^2}{17.5} + \dfrac{(31 - 17.5)^2}{17.5}$

$\chi^2 = 3.87 + 3.87 + 4.97 + 4.97 + 0.03 + 0.03 + 10.41 + 10.41 = 38.56$

(c) Degrees of freedom $= (R - 1) \times (K - 1)$

df $= (4 - 1) \times (2 - 1) = 3$ degrees of freedom

Table 19

Critical value: 0.05		Critical value: 0.01	
Degrees of freedom	Value	Degrees of freedom	Value
1	3.84	1	6.63
2	5.99	2	9.21
3	7.82	3	11.30
4	9.49	4	13.30
5	11.10	5	15.10

Reading from the critical values table it can seen that the calculated value of χ^2 is higher than the critical value and the probability that the result occurred by chance is less than 99%.

(d) To summarise, using the chi-squared test it has been shown that there is a significant difference in the degree of pebble roundness between the upstream and downstream sites.

Mann–Whitney U test

(20) (a)

Table 20

Upstream		Downstream	
x	Rank (r_x)	y	Rank (r_y)
15	19	4	1
8	7.5	8	7.5
22	24	10	10
32	27.5	6	4
16	20	19	22
18.5	21	14	18
34	29	6	4
32	27.5	13.5	16.5
19.5	23	7	6
13.5	16.5	5	2
28	26	12.5	13
10.5	11	12	12
13	14.5	8.5	9
24.5	25	6	4
45	30	13	14.5
No. in sample (N_x) = 15		No. in sample (N_y) = 15	
Total of rank scores (Σr_x) = 321.5		Total of rank scores (Σr_y) = 143.5	

(b) Upstream site (x):

$$U_x = (15 \times 15) + \left(\frac{15(15 + 1)}{2}\right) - 321.5$$

$U_x = 23.5$ (smaller value)

Downstream site (y):

$$U_y = (15 \times 15) + \left(\frac{15(15 + 1)}{2}\right) - 143.5$$

$U_y = 201.5$

(c) Using the smaller value (23.5) we can now test the result for significance using the critical values table (Table 21). In this test both sample sizes are the same, 15.

Table 21 Critical values for the Mann–Whitney U test

Size of the largest sample (n_2)

Size of the smallest sample (n_1)	5	6	7	8	9	10	11	12	13	14	15	16	17	18	19	20	21	22	23	24	25	26	27	28	29	30
3	0	1	1	2	2	3	3	4	4	5	5	6	6	7	7	8	8	9	9	10	10	11	11	12	13	13
4	1	2	3	4	4	5	6	7	8	9	10	11	11	12	13	14	15	16	17	17	18	19	20	21	22	23
5	2	3	5	6	7	8	9	11	12	13	14	15	17	18	19	20	22	23	24	25	27	28	29	30	32	33
6		5	6	8	10	11	13	14	16	17	19	21	22	24	25	27	29	30	32	33	35	37	38	40	42	43
7			8	10	12	14	16	18	20	22	24	26	28	30	32	34	36	38	40	42	44	46	48	50	52	54
8				13	15	17	19	22	24	26	29	31	34	36	38	41	43	45	48	50	53	55	57	60	62	65
9					17	20	23	26	28	31	34	37	39	42	45	48	50	53	56	59	62	64	67	70	73	76
10						23	26	29	33	36	39	42	45	48	52	55	58	61	64	67	71	74	77	80	83	87
11							30	33	37	40	44	47	51	55	58	62	65	69	73	76	80	83	87	90	94	98
12								37	41	45	49	53	57	61	65	69	73	77	81	85	89	93	97	101	105	109
13									45	50	54	59	63	67	72	76	80	85	89	94	98	102	107	111	116	120
14										55	59	64	67	74	78	83	88	93	98	102	107	112	118	122	127	131
15											64	70	75	80	85	90	96	101	106	111	117	122	125	132	138	143
16												75	81	86	92	98	103	109	115	120	126	132	138	143	149	154
17													87	93	99	105	111	117	123	129	135	141	147	154	160	166
18														99	106	112	119	125	132	138	145	151	158	164	171	177
19															113	119	126	133	140	147	154	161	168	175	182	189
20																127	134	141	149	156	163	171	178	186	193	200
21																	142	150	157	165	173	181	188	196	204	212
22																		158	166	174	182	191	199	207	215	223
23																			175	183	192	200	209	218	226	235
24																				192	201	210	219	228	238	247
25																					211	220	230	239	249	258
26																						230	240	250	260	270
27																							250	261	271	282
28																								272	282	293
29																									294	305
30																										317

NB: This page is also available as a pdf download from **www.saburchill.com/IBbiology/downloads/002.pdf**
Level of significance: 5% ($P = 0.05$)

This table shows that the critical value at the 0.05 level of significance is 64. As the smaller U value of 23.5 is lower than the critical value, the null hypothesis can be rejected. We can be 95% certain that the result obtained did not occur by chance and that there is a significant difference in bedload size between the upstream and downstream sites.

(d) The bedload was expected to be smaller in size at the downstream location for a number of reasons. First, upper-course streams generally have a lower discharge and competence than rivers in a downstream, lower-course location. Therefore, larger bedload material is seldom transported from the upper course, near the source. Additionally, material which is transported downstream by traction and saltation is worn down by erosion processes, such as attrition.

Fieldwork

This section provides:

- General guidance on the nature of the fieldwork to be undertaken at AS and A2. You are advised to read this section (pages 49–54) in its entirety before embarking on any fieldwork.
- Exemplars of the types of question you could be asked in each of the examinations for Unit 2 and Unit 4A. Make sure you understand the difference in the demands of the two types of question.
- Three examples of fieldwork undertaken by students.
- Examples of examination questions on fieldwork and the answers to them, for Unit 2 and Unit 4A, with commentary.

Assessment of fieldwork at AS and A2

At AS and A2 you are required to undertake investigative work in the field. This allows you to develop skills associated with planning investigations, collection of primary and secondary data, and presentation, interpretation and evaluation of results. In Year 12 you are expected to take the Unit 2 examination (1 hour) where this fieldwork is assessed in Question 2.

At A2, in Year 13, you are required to sit either Unit 4A — an examination consisting of structured and extended questions based on your own fieldwork investigation (1½ hours), or Unit 4B — an examination of structured and extended questions, including research-based skills questions, based on an advance information booklet, which is pre-released (1½ hours). Both papers carry 60 marks.

This section provides advice on the piece of fieldwork you should carry out for Unit 2 and for Unit 4A Section A, where it is assumed that you have undertaken a complete fieldwork activity, including writing it up. The personal fieldwork investigation that you undertake has this broad task:

> the individual investigation of a geographical argument, assertion, hypothesis, issue or problem

At AS the only restriction is that the fieldwork must relate to the content of Unit 1 (Rivers, floods and management, Population change, Cold environments, Coastal environments, Hot desert environments and their margins, Food supply issues, Energy issues and Health issues).

At A2 there are no restrictions on the topic studied other than it should be geographical and include primary, and where relevant secondary, data collection, should be based on a small area of study and be linked to the content of the specification. This is a very wide brief. It is also clear that this investigation/fieldwork must be completed at a level above that done at AS.

Aim, research questions and hypotheses

The **aim** of your project is what you are generally trying to achieve in your fieldwork location. This will depend on time, environmental conditions, equipment available and risk assessments. For example, your aim might be:

> To study the changes in river characteristics downstream

Alternatively, you might want to express what you are investigating as a **research question**:

> Which factors influence channel characteristics on a small stretch of a local stream?

What are the differences in the characteristics of a climatic climax vegetation and a plagioclimax vegetation?

You could also think about breaking down the overall question into smaller sub-questions. Go for two or three such questions, such as:

What management strategies have been used to manage coastal erosion at Y?

How effective have previous management strategies been at Y?

What are the attitudes of people to the management strategies used at Y?

A **hypothesis** is a statement based on a question, such as:

A number of factors cause flooding to occur at P.

A range of management strategies are used to protect area P from flooding.

Not everyone thinks the flood management strategies at P are effective.

You may wish to **evaluate an issue** in a local area. For example:

What are the different attitudes of local people to the plan to build a Tesco Local store in J?

What are the causes and consequences of the increase in student accommodation in suburb K?

Is the proposed development of new housing in the area of Manor Farm necessary?

It is important early on to start collecting resources about both the specific area and the general theme chosen. Note that for each of these you should be aware of the concepts or processes (sometimes referred to as the **underpinning theory**) that led to the idea for the fieldwork in the first place. Make sure you have done research into these before beginning the exercise.

Risk assessments are important too. Make sure what you do is safe, and that others know where you will be. Be fully aware of what you have done to be safe — you may be set questions on this.

Methodology and data collection

This part of the project is where you sort out the most suitable techniques to meet your aim and investigate your hypotheses or research questions. You therefore need to make decisions about sampling.

Many investigations rely on a representative sample from the parent population. This population may, for example, be pebbles on a beach, traffic on a road or residents in an area. All samples should be proportional to the size of the total population. Sampling may be random, systematic, stratified, clustered or a combination of these. It is important to be able to justify the decisions you make about sampling. Consider using a pilot study.

When you are describing your methodology, consider using tables to summarise your work. Tables 1 and 2 are examples based on collecting primary data (i.e. data collected by you) and secondary data respectively. You should also indicate which techniques were carried out on your own and which as part of a group.

Table 1 Primary data collection

Technique	Why used/ purpose	Method: when/ where	Justification of sampling type	Problems/ limitations	Improvements	Methods to be used for analysis
River surveys						
Land-use survey						
Field sketches						
Photographs						
Questionnaire						

Table 2 Secondary data collection

Technique	Why used/ purpose	Method: when/ where	Justification of sampling type	Problems/ limitations	Improvements	Methods to be used for analysis
Government statistics						
Local area plan						
Local newspaper						
Websites						

You could include neat hand-drawn sketches and maps in this part of the report, as well as labelled photographs to show how any complex equipment was used and how you maintained reliability and accuracy. If you are using a questionnaire, for example, consider including an annotated blank form to show the reasons why you chose those questions and why you put them in the order in which they appear.

Data presentation

The purpose of this part of the report is to present the data you collected in a way that is easy for the reader to understand. Interpretation may include looking for trends and patterns as well as trying to spot anomalies. Try to:

- include a wide range of appropriately chosen representation techniques
- decide whether the data need spatial (i.e. mappable) techniques or non-spatial techniques such as pie graphs
- include photographs, which are useful to any study as long as they are annotated
- be wary of using computer graphics that look impressive but which may be useless at showing patterns or trends. Hand-drawn maps and diagrams may be better than 'death by bar chart'
- go for a range of quantitative (numerical) and qualitative (e.g. viewpoints, field sketches) representations
- use a spell check (English UK not English US version) but do not rely on it — ask a friend to proofread your work

The data presentation section does not have to be isolated — often it is best integrated into an analysis or results section. However, you should be clear in your mind which of your techniques involve *presentation* and which *analysis*.

Results and analysis

This is often the hardest but most important part of the investigation. There are several steps involved:

- Collate data, exchanging information across the group if you are not carrying out individual data collection. You could enter the data into a spreadsheet or database program.
- Describe data in terms of trends, relationships and correlations. Always refer back to your original aim, research questions and hypotheses.
- Carry out statistical analysis. The main point to remember is to find and write about geographical aspects of your investigation rather than get lost in numbers. There are several statistical exercises that are available — make sure you select the most appropriate technique, complete it accurately, and be sure what the final result actually means and how significant it is.
- Explain your findings. Refer back to the research questions or hypotheses in the introduction. You should start looking for and exploring relationships between the data. How do the results measure up to expectations?

Conclusion and evaluation

At this stage you should:
- return to your original aim, research questions or hypotheses
- draw together all the threads of your analysis section
- evaluate the success of the data collection techniques, area of study chosen etc.
- suggest other avenues of enquiry
- suggest a possible future for the area studied based on your understanding. This might take the form of a new management plan or a map of your suggested solutions to a problem
- evaluate the strengths and weaknesses of your fieldwork investigation. Consider using a strengths, weaknesses, opportunities and threats (SWOT) table

Writing up

You may be asked by your teacher to write up your completed enquiry for the examination that will follow. You may also want to submit it as part of the Extended Project Qualification (EPQ). Here are some hints and tips for this process:
- Keep all project work together in a separate folder. Organise this into sections for easy retrieval before discussions with your teacher.
- Organise your work under the headings given above.
- The report should be word processed if possible and well organised.
- An appendix section may be useful for including evidence, raw data or supplementary resources. Don't make it too long or include important information that should be in the main part of the report.
- There should be a bibliography at the end of the project that lists all the resources you have used. Ideally, you should have used a range of media, e.g. books, video/DVD, the internet, magazines and newspapers. Remember that any diagrams or text you have copied from secondary sources must be acknowledged.

AS Unit 2

For AS Unit 2, the fieldwork plan in Table 3 may help in collating your thoughts.

Table 3 Fieldwork plan

Stage	Points to consider	Comments on tasks/ methods to be used
Aim	What is the aim of the study?	
	What are you trying to find out?	
	What is the underpinning theory?	

Stage	Points to consider	Comments on tasks/ methods to be used
Hypotheses	What hypotheses can be tested?	
	Which variable can be correlated against another or in relation to some other variable such as distance?	
Data collection	What methods can be used to collect data for each of your hypotheses?	
	Why are they the best?	
	What equipment will be required to carry out the investigation?	
	What are the risks associated, and how do you respond to them?	
Data presentation	What method(s) would be suitable to test relationships between the variables being measured?	
	Why is this method suitable?	
	Illustrate this technique	
Data analysis	What method(s) could be applied to analyse the data?	
	Is there any correlation between the variables?	
	What are the advantages of using this method?	
Results	What did you learn?	
	Were there any anomalous results?	
	What/how did it contribute to your understanding?	
Improvements	What could you do to improve it?	

Some of the questions that could be asked In Unit 2 Question 2 are given in Table 4, divided into the five main elements of enquiry.

Table 4

	Starting point	Data collection	Skills of presentation	Interpretation	Conclusions/evaluation
1	Outline the aim and describe the theory, idea or concept from which your aim was derived	Outline and justify one method of data collection that you used	Describe one method used to present your data	What are the advantages and disadvantages of the analysis technique(s) that you used?	How far did your fieldwork conclusions match the geographical theory, concept or idea on which your study was based?
2	Explain the geographical concept, process or theory that underpinned your enquiry	Examine the limitations of your chosen methodology	Describe one application of ICT skills in carrying out your fieldwork and comment on its usefulness	Outline and justify the use of one or more techniques used to statistically analyse your results	Summarise your findings and suggest how this enquiry could be improved
3	Outline one source of information that you used and assess the extent to which it was 'fit for purpose'	Outline one hypothesis and describe one methodology for primary data collection in relation to this	Describe and illustrate one technique you used to present data in this enquiry	Name one technique of data analysis and describe how it was used	Making specific reference to your results, suggest how your enquiry could be improved
4	Explain how you devised your aim and how you responded to the risks associated with your chosen site for fieldwork	How did you respond to risks associated with undertaking primary data collection	What difficulties did you face when presenting your results?	What is meant by the term 'significance' in the analysis of fieldwork data?	In what ways would your conclusions be of use to other people?
5	Describe the location of your fieldwork and explain why it was suitable for your investigation	Discuss the strengths and weaknesses of the method of data collection	Describe a method of presentation that you used in your investigation and indicate how the chosen method was useful	In the context of the analysis of fieldwork data, outline the meaning of 'anomalies'	Drawing upon your findings, explain how your enquiry improved your understanding of the topic area

A2 Unit 4A

The AQA specification states the following regarding Unit 4A:

> The examination will test candidates' knowledge and understanding of the subject matter relating to their investigation and its links with other aspects of geography; their critical approach to the methodology, approaches and techniques associated with data gathering, presentation and analysis; of the findings of the investigation and its contribution to furthering candidates' geographical understanding and that of the role of fieldwork inquiry in geographical study.

Unit 4A is a more challenging examination than AS Unit 2. The questions have to demonstrate elements of both 'synopticity' and 'stretch and challenge'. Essentially, the differences between the two examinations can be summarised as:

- Unit 2 assesses what I did on my fieldwork
- Unit 4A assesses why I did what I did on my fieldwork

Therefore, all questions in Unit 4A will be evaluative. Again there is a limited range of questions that could be asked — Table 5 provides a summary of possible alternatives.

Table 5

	Starting point	Methods	Skills	Interpretation	Conclusions/evaluation
1	In the light of your aim why was the location selected?	Describe and justify the risk assessments that you did	Data presentation skills — why selected and what were the strengths and weaknesses?	How and why were the techniques of interpretation useful in developing understanding?	Evaluate the success of the investigation in the light of the aim
2	Why was the aim or hypothesis chosen?	Justify the methods used to collect data	Data presentation — evaluate the alternatives	How did the aim/data influence the skills of interpretation used?	What did you learn from this study — geographical understanding?
3	How important was the theory in the choice of aim?	Explain how and why data were selected for the investigation	What skills were used to present the data and why were they useful?	Interpret the results in the light of the aim — how was understanding broadened/contributed to?	How did theory influence your evaluation of this investigation?
4	How important was location in the choice of aim?	How were data selected? What improvements would you make?	How did the aim influence presentation methods?	How did the results influence interpretation?	How can your investigation be further developed and extended?
5	How important was theory in the choice of location?	Explain the importance of sampling to your investigation	How did results influence presentation skills?	How important were anomalies in the interpretation of the data collected?	In what ways would your conclusions be of use to other geographers?

The next section gives extracts from three enquiries that may assist you in preparing for, or writing up, your own fieldwork.

Example fieldwork enquiries
Enquiry 1: Rivers based

Extracts from the write-up of a fieldwork enquiry based on a river are given. The sections in italics are the words used by the student.

The context comes from the AS course, but both AS and A2 skills are used.

Hypotheses
1   *The size of bedload (clasts) decreases downstream.*
2   *The shape of bedload (clasts) changes downstream.*

Location
A detailed location was given of the study area (in northwestern England), both in terms of maps and written description.

Characteristics of the area
The river studied flows off the western side of the Pennines. The main underlying rock type is sandstone, with scattered bands of shale, both of which are impermeable. The rivers often have large volumes of water flowing in them which bring clasts into the river. The surrounding area is an upland plateau with steep v-shaped valleys cut into the landscape by the rivers. The steep slopes create interlocking spurs as the narrow channel of the valley winds downstream. The steep gradient of the valley sides could also increase the rate at which clasts are brought into the river.

The area has characteristically low temperatures and high rainfall. The vegetation of the area is moorland with rough grasses and heather on the slopes, and peat on the hill tops.

The course of the river is disrupted by human activity, a water treatment plant, which could affect the movement of clasts downstream. In addition, the vegetation is burned on a regular and controlled basis by sectioning to give ideal breeding conditions for grouse. This could affect rates of surface runoff and the hill slope transportation processes.

Relevant geographical theory
Brief sections were given on each of the following:
  * the origin of the clasts: a discussion of relevant river erosional processes (hydraulic action, abrasion, solution, attrition, bank slumping and turbulence)
  * weathering: physical, chemical, biological
  * hillslope transportation processes: creep, flows, slides and falls
  * long profile of a river
  * vertical erosion and lateral erosion by a river
  * river transportation processes: traction, saltation, suspension, solution
  * the Hjulström curve

Methodology

I used a method of systematic sampling to select ten sites along the river from which to take data. Ten sites is the minimum number to be used in a Spearman rank test, yet also provided a sufficient spread of data. Systematic sampling ensures the sites are spaced equidistantly along the river allowing change to be seen. Random sampling could lead to the possibility of a majority of sites in one section of the river, for example in the upper course, which would not be useful in proving/disproving the theory that clasts will decrease in size and become more rounded downstream.

However, areas at these specified places could not always be accessed as they were dangerous to reach (e.g. Site 1). In addition, Site 6 had to be moved further downstream (1,000 m from Site 5) as the water treatment plant created an obstruction, and Site 4 was moved slightly upstream (only 600 m from Site 3) because of the large meander in the river — an obvious geographical feature which could significantly affect the results.

Once at each site I used a method of random sampling to identify where to obtain measurements from. I laid a tape measure along the river bank for 5 m, picked a random number from the random number table, and used this to mark the position in the river where I would take measurements from. This method was designed to get rid of bias and avoid me choosing to measure the easiest section of the river, which might not necessarily be representative of the area due to increased erosion from others disrupting it.

Identifying bedload

It would be time consuming and unnecessary to measure every clast in the river, so I chose to record measurements from a sample of 30 clasts at each site, which is a big enough value to be reflective of the area, and is also the minimum sample size of chi-squared and Mann–Whitney U statistical tests. Using a larger number of clasts also reduces percentage error — if only ten rocks were measured and one was anomalous then 10% of the results would be anomalous, but with 30 clasts this would be only 3.3%, a more acceptable level of error.

To choose which clasts to measure I used a random systematic sampling technique. Firstly I measured the width of the river channel and divided into tenths to give ten positions from which I would measure three clasts at each (one from that point, one from 1 m upstream, and one from 1 m downstream of the point). It was important to be systematic as clasts would be sorted differently at different points in the river. However, the method was random in that I used a random number table to identify a number within the value of the first 1/10th of the width which was used as a start point from which to measure each further 1/10th from. It was important to be random to avoid simply choosing a convenient point at which the rocks would be easiest to measure.

Measuring the bedload

Size

I recorded the volume of each clast by measuring length, width and depth and multiplying them:

$$Volume = length \times width \times depth$$

However, the main problem with this method of calculating volume is that angular and rounded clasts will appear to have the same volume in my results, when in fact this is not the case. This problem will be further exaggerated downstream as theoretically rocks here become more rounded, resulting in a larger difference between apparent and actual volume.

Shape

I recorded each clast in one of the following categories:
- angular (A): angular corners, angular edges, flat faces
- sub-angular (SA): rounded corners, angular edges, flat faces
- sub-rounded (SR): rounded corners, rounded edges, flat faces
- rounded (R): rounded corners, rounded edges, rounded faces
- sand/silt/clay (N): no visible shape, 0.1 cm^2

The main problem with this method is that it is qualitative, and relied on my own judgement being consistent throughout. However, there is no clear way to assess shape — it is subjective and therefore very hard to distinguish between sub-angular and sub-rounded rocks which could give inaccurate patterns when analysing my results. I can only hope that I have judged each rock to a similar standard.

Gradient of the river channel

Gradient of the river affects the velocity of the water — if steep, water will move more quickly and there is increased transportation. I measured the gradient by holding two poles 5 m apart, on the surface of the water, in the middle of the width of the river. I then used a clinometer to calculate the angle of the slope.

I used the clinometer to measure the angle of the slope on the valley side. Steep slopes are evidence of weathering, for example freeze–thaw. This factor affects where the clasts originate from by causing increased surface runoff, and affects the size and shape of the clasts as they enter the river quickly before further weathering. In addition I recorded whether there was a floodplain at the site, and if so, measured the size in 1 m paces. This could have the opposite effect to a steep valley side, as clasts remain stationary and are weathered.

While observing the river I looked for any significant geographical features, for example, at Site 2, where there was a waterfall, causing the river bed to be broken up. This could affect clast size and shape from increased erosion by hydraulic action, hence could create a possible anomaly in my results. I observed any biological weathering, such as trees growing into the rock and dense vegetation, for example at Site 1. This could hold up transportation of clasts into the stream. On the other hand, land could be bare which would speed up transportation processes. Any human intervention, for example man-made walls, was also observed and recorded as this could affect river/clast size and transportation, hence affecting whether the site fits in with theory. A significant example of this was at Site 6, where a large water treatment plant disrupted the flow of the river.

Results and analysis

One of the major problems that I found in terms of presenting the data was the huge range of clast sizes. There were extremely few large clasts, one was over 190,000 cm³ at Site 6, whereas the vast majority of clasts were smaller than 20,000 cm³. There were also some large clasts at Site 9. To overcome this I used a semi-logarithmic dispersion graph (Figure 1), with clast size on the vertical semi-log axis, and distance downstream on the horizontal axis. One advantage of this graph is that it allows a clearer pattern of the dispersion points to be shown. Another advantage of this graph is that it allows individual points to be distinguished more easily, and condenses the few points near the top of the graph that would otherwise be extremely spread out.

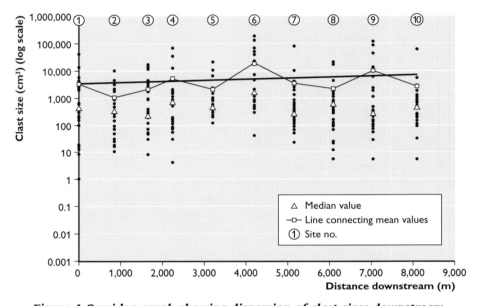

Figure 1 Semi-log graph showing dispersion of clast sizes downstream

The trend line on the graph showed an unexpected result — that clast sizes actually increased downstream from Site 1 to Site 10. The mean fluctuates from site to site showing no obvious pattern, as did the median. The main problem with this graph is that bigger anomalous values appear to be closer to the trend line due to the uneven scale, but this can be overcome with accurate analysis of the graph.

I have also drawn a scattergraph (Figure 2) to show the relationship between the distance downstream and each of the mean values at each site. Again, the graph shows a weak positive correlation, that clast sizes increase downstream. This pattern contradicts classic theory. However, the obvious anomaly in the whole pattern of things is that the clast sizes at Site 6 are so high. I have drawn a graph (Figure 3) of this relationship from Site 6 to Site 10 and you will see that the relationship becomes negative (despite an anomaly at Site 9).

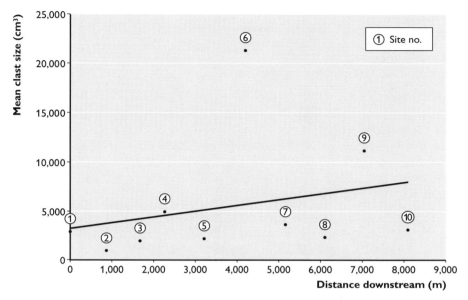

Figure 2 Graph to show mean changes in clast size with distance downstream

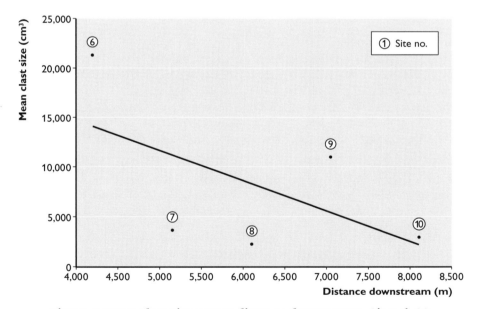

Figure 3 Mean clast size versus distance downstream, Sites 6–10

I decided to divide the sites at Site 6 as this seemed to be the most obvious anomalous result that might cause the negative correlation — downstream from here in the lower course of the river the data correspond with classic theory. This is explainable by the presence of an obvious man-made interruption to the river just prior to this site — the water treatment centre.

Statistical analysis
I performed two statistical tests on the data.

Spearman rank
I tested a statistical relationship between the distance downstream and the mean clast size. I made two hypotheses:
- *the null hypothesis: there is no significant correlation between mean clast size and distance downstream*
- *the alternative hypothesis: there is a significant positive correlation between mean clast size and distance downstream*

If the value calculated by the Spearman rank test is greater than or equal to the critical value, the null hypothesis is rejected and the alternative hypothesis adopted. If the calculated value is less than the critical value the null hypothesis is retained.

The result of the test was 0.18 which is less than the 5% critical value of 0.56 so that proves that there was no statistical positive correlation between mean clast size and distance downstream. This contradicts the seemingly positive relationship shown in the scattergraph. The test shows that my data are not significant evidence to disprove the theory that clast size decreases with distance downstream.

Mann–Whitney U test
This tests the significance of the differences between the size of the clasts at two sites — I have chosen Sites 1 and 10 as these are furthest apart and so should theoretically show the most significant difference. I made the following hypotheses:
- *the null hypothesis: there is no significant difference between clast sizes at Sites 1 and 10*
- *the alternative hypothesis: there is a significant difference between clast sizes at Sites 1 and 10*

If the calculated value for U is greater than or equal to the critical value for U, the null hypothesis is rejected and the alternative hypothesis adopted. If the calculated value for U is less than the critical value for U, the null hypothesis is retained.

The calculated value for U was 471.5 and the critical value for U is 582.6, so the null hypothesis is adopted — there is no significant difference between clast sizes at Sites 1 and 10.

Previous graphs have shown that from Sites 1 to 5 the river does not accord with the theory. However, from Sites 6 to 10 the river follows classic theory as mean clast size decreases with distance downsteam. Therefore I repeated the Mann–Whitney U test using data from Sites 6 to 10 to see if the same changes can be seen. When calculated, U is 679, and this is greater than the critical value for U at the 0.05 and 0.01 significance level (673.2). Hence I can conclude that although as a whole the river does not accord with classic theory, it does show a significant negative correlation between clast size and distance downstream from Sites 6 to 10.

Clast shape analysis

According to classic theory clasts should become less angular and more rounded further downstream. As the load is transported in the river, processes such as attrition knock off the angular edges forming smoother, more rounded clasts. I have constructed a divided bar graph (Figure 4) to show the variation in clast shape from Sites 1 to 10 using the data from the 30 rocks I classified at each site.

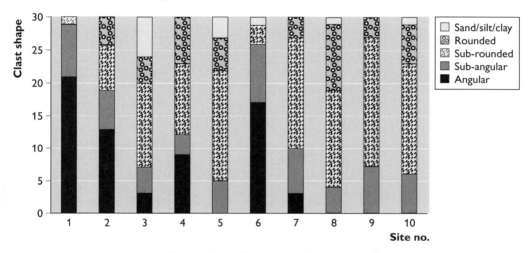

Figure 4 Graph to show clast shape at each site

The graph shows the general pattern to be that the number of sub-angular and angular rocks decreases downstream, while the number of rounded and sub-rounded clasts increases. This is in accordance with the theory. At Site 1, 70% of the clasts are angular with another 27% as sub-angular — the remaining clasts being sub-rounded with none at all classed as rounded. This is the highest proportion of angular/sub-angular clasts found at any of the sites, and so fits well with the theory.

Rocks do generally become more rounded downstream — from Site 8 onwards none of the clasts measured were classified as angular, and only 20% sub-angular by Site 10. On the other hand this is a higher percentage of rocks in the angular/sub-angular category than at Site 5. At both Sites 9 and 10, 77% of all clasts were found to be rounded or sub-rounded. However the highest percentage of sub-rounded and rounded clasts was found at Site 8 (83%) so although the river does generally accord with theory, this is not perfect as the highest percentage should be found at Site 10.

The main anomaly appears to be Site 6 where 57% of the clasts are angular. A further 23% of clasts at this site are classed as sub-angular, only 10% as sub-rounded and none as rounded. This again could be due to the presence of the water treatment works at this section of the river. Site 3 also shows as a slight anomaly with very few clasts classed as either angular or sub-angular, and 20% of all clasts being classified as sand or no shape.

Statistical analysis

Chi-squared tests the significance of differences between the shapes of the clasts at two sites. As the most significant change should theoretically be seen between Sites 1 and 10, as they are the furthest apart, I have used data from these two sites.

I made two hypotheses:

- *the null hypothesis: there is no significant difference between clast shape at Sites 1 and 10*
- *the alternative hypothesis: there is a significant difference between clast shapes at Sites 1 and 10*

If the calculated chi-squared value is greater than or equal to the critical chi-squared value then the null hypothesis is rejected, and the alternative hypothesis is adopted. However, if the calculated chi-squared value is less than the critical value, then the null hypothesis is retained.

As the calculated chi-squared value is 41.5 and the critical value is 11.34 at the 0.01 significance level the alternative hypothesis is adopted and the null hypothesis rejected. Therefore clast shapes did change and become more rounded downstream due to erosional processes such as attrition.

Conclusions

The aim of my investigation was to test the classic river theory relating to bedload (clast) size and shape, namely that the size of the bedload decreases downstream and that the bedload becomes more rounded downstream.

I have found that for this river as a whole there is an increase in clast size over the ten sites which does not accord with the theory, although the Spearman rank statistical test has shown that the positive correlation is of no significance and is not strong enough to disprove the theory. Similarly by using the Mann–Whitney U test I have shown that there is no significant difference in clast size at Sites 1 and 10 for this river.

This can be explained by the unusual character of the upper course, where at Site 4 I found meanders cutting into glacial material and bringing new large rocks into the river channel. In addition there is a steep valley side, making hill slope transportation processes effective here and delivering a high proportion of large clasts to the channel. The expected trend of the theory was also interrupted by man — Site 6 where the site of a water treatment works lay just upstream. Here the size of the clasts was unusually large. This may be due to the fact that as the water is slowed for extraction, sediment falls and it is then bulldozed out of the river. This is necessary to allow for the smooth flow of water into the extractors. Below this the river effectively starts again and receives its load from the valley sides — an influx of new large and angular rocks from the sides.

Despite the contradiction to the theory in regards to bedload size, the river follows theory in that generally bedload becomes more rounded downstream. This is due to erosion of the clasts, particularly by attrition (where clasts in the river channel bang against each other, knocking off angular edges and making them more rounded downstream). The clasts originate in the upper course where the valley sides are steep and weathering processes cause rocks to become loose from the valley sides and fall by hill slope transportation processes into the river channel. These clasts have therefore had little time to be eroded by attrition and so are angular.

The main anomaly in terms of clast shape was at Site 6 where clasts were found to be uncharacteristically angular. Again this is explained by the presence of the water treatment works at this site which disrupts the river by removing the eroded rounded clasts that originated in the upper course. Similarly to the clast size, the river effectively starts again at this site with new angular clasts originating from the steep valley side by rock falls and slides.

I have concluded that generally theory does apply to the river I studied, particularly in terms of bedload shape, although bedload size was significantly disrupted by man-made influences. Could this mean that classic theory is simplistic? This is also reinforced by natural factors — there are certain places where the input of certain types of load is due to increased amounts of glacial deposits.

Evaluation

I chose ten sites by systematic sampling to ensure a sufficient spread of data and allowing a change to be seen along the river by placing them equidistantly. However, Site 1 was not the source of the river as this would have been too dangerous to reach. In addition Site 6 was moved downstream as the water treatment plant created an obstruction. Also at Site 4 there was an obvious geographical feature which would have affected the results greatly (a large meander) so this site was moved upstream. There is a strong possibility that these changes have affected my results.

I took a sample of 30 clasts at each site in the hope that this would be representative of the area, and also to allow statistical testing when analysing. By taking a bigger sample my results would have been made more reliable although this would have been time-consuming, making it very impractical. The size of the clast was determined by measuring the width, length and depth — the main problem with this being that angular and rounded clasts may have the same dimensions and therefore have the same volume. Yet, rounded clasts would be smaller. Measuring the rock accurately with a ruler was also difficult as the sides of the rock are not straight and lengths vary across the clast. Alternative methods would have included a water displacement test, but this would have been impractical to transport and lifting each bulky rock from the river bed into the water tank would have proved difficult. In addition I encountered problems with deciding shape as this is a subjective test, relying solely on my own judgement being consistent throughout with no distinct line between a sub-angular and a sub-rounded rock.

In addition the length of the river I studied was only 10 km. The river may have not flowed far enough to have an impact on size and shape. Rocks further downstream may have not undergone enough attrition to be affected.

The only major problem found with my results was the incredibly anomalous value of Site 6. However, this was undoubtedly caused by the influence of the man-made water treatment works just upstream from where the measurements were taken. The smaller anomaly of Site 4 was also explainable by the obvious geographical feature (the meandering of the river in this area and the deposition of glacial material) which will have influenced the data.

I used Spearman rank statistical testing to evaluate the relationship between clast size and distance downstream. In some respects this is a good test as it takes into account data from all the sites. The problem with it was that, though it used the mean value of rock size, a few extremely large clasts still had a big effect. On the other hand I also used the Mann–Whitney U test to see if there was a significant difference between clasts at Sites 1 and 10. This test considers all the clasts at each of the two sites examined rather than taking an average. This increases confidence in the results as any anomalous results are identified individually and do not affect how large or small the other clasts appear to be. However, the technique takes no note of the data from the other 8 sites and works best when sites are spaced far apart from each other as otherwise changes in clasts are too small for the test to recognise. This weakness means that the fact that the test proved positive for the data between Sites 6 and 10 is less significant. It would have been far more significant if both tests (between Sites 1 and 10, and Sites 6 and 10) had proved positive, but this was not the case.

When evaluating the significance of change in clast shape downstream I used the chi-squared test which was successful in showing that my data support the theory. Despite this the strength of this evidence to prove that clasts become more rounded downstream could be weakened. The problem with this test is that it only compares two sites at a time. It does not show that clast size decreases gradually downstream, only an overall decrease, whereas theory states that a higher proportion of clasts should become rounded at each site.

Overall my conclusions are not strong, and by using more accurate methodology (however impractical), my results would have been more accurate and ultimately my conclusion more reliable. In addition it would be useful to repeat this investigation at different times of the year in varying conditions. For example after periods of heavy rain and high discharge there may have been more erosion and rounding of the clasts. Also in winter there would be more expanding and contraction of the valley sides and more mass movement leading to more large and angular rocks in the bed of the river.

My study is just one that could be carried out on a number of rivers, and this has to occur to see if the same pattern of results is found in them. By investigating different rivers I would be able to make a more accurate and reliable judgement on the validity of classic theory compared to the result I have found from this investigation.

Final comments

You might want to review this piece of fieldwork. You could ask yourself how it could have been improved further.

Enquiry 2: Ethnic segregation

Extracts from the write-up of a fieldwork enquiry based on ethnic segregation are given. The sections in italics are the words used by the student, except that the area names have been removed.

The context arises from the A2 course, and a wide range of skills is used.

Hypotheses

1   *The town of X exhibits characteristics of ethnic segregation.*
2   *This degree of segregation is evident in the built environment.*

Location

A detailed location was given of the study area (in northwestern England), both in terms of maps and written description.

Characteristics of the area

Town X (Figure 5) has a total population of over 200,000. Its main industry in the past was the cotton industry. This called for the import of cheap labour from the east, and as a result the town's population exploded and has been rising ever since, including the numbers of ethnic minorities. This study is intended to investigate the degree of ethnic segregation in the town, and to examine the signs of this segregation within the built environment. I shall use census data and fieldwork research to identify both of these features.

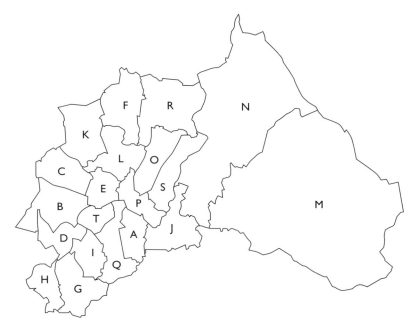

Figure 5 Town X — wards

Relevant geographical theory

No further geographical theory was given. The student embarked on the enquiry straightaway.

Methodology

For hypothesis 1, I used mostly census data, via the internet. The data were then processed into relevant categories. I classified people of Bangladeshi, Indian and Pakistani origin as New Commonwealth Immigrants (NCI) (see Tables 6 and 7). This allowed tabulation and initial analysis. Data from the census were then mapped using a variety of maps to determine if a zone of ethnic segregation was evident and to see if this concurred with the statistics that might indicate forms of urban deprivation. I also plotted a Lorenz curve to show the distribution of people in the town.

This type of analysis based on secondary data is limited by the nature of the source. In particular, the data are over 5 years out of date. Also the use of wards is not ideal as there can be variations within their boundaries. Each ward can have a mixture of types of area within them. Some wards are very small in area, whereas others are very large (see Figure 5).

Table 6 Selected census data (by ward), 2001

Ward	% of ward population that are NCI	% unemployed	Average household income (£ 000)
A	26.6	6.1	19.3
B	1.6	2.4	27.3
C	3.2	2.5	27
D	2	3	23.7
E	52.9	7.2	16.4
F	0.8	2.1	27.9
G	1.5	2.8	27.2
H	1.2	3.3	23
I	4	4.8	20.4
J	2.6	4.5	30.9
K	0.7	2.2	27.5
L	1.3	2.6	27.9
M	0.7	1.9	33.6
N	0.5	2	34.7
O	2.2	5.2	23.6
P	36.6	6.2	18.9
Q	24.6	4.3	20.7
R	3.2	2.8	27.3
S	3.3	3.2	24.3
T	53.7	6.5	19.8
Town X	**11.5**	**3.7**	**25.1**

Table 7 Selected census data (by ward), 2001

Ward	No. of NCIs in the ward	% of NCIs	% area of the borough
A	2,972	11.8	1.3
B	181	0.7	2.7
C	340	1.4	2.3
D	199	0.8	1.7
E	6,310	25.1	1.7
F	87	0.4	3
G	166	0.7	3.3
H	124	0.5	1.3
I	396	1.6	2
J	260	1	3.7
K	78	0.3	3.3
L	131	0.5	2.3
M	98	0.4	36.5
N	56	0.2	20.1
O	208	0.8	2.3
P	3,942	15.7	1.3
Q	2,589	10.3	2.7
R	335	1.3	4.7
S	436	1.7	2.7
T	6,225	24.8	1
Town X	25,133	100	100

I tested hypothesis 2 by means of a field environmental survey. An assessment booking sheet was devised based on examples found in textbooks (Table 8). This approach measured things such as landscape quality, amount of derelict land, vandalism, industrial premises, noise, air pollution, access to public space and shops and services, and traffic flows. The scoring system operated on a points basis, with low-quality areas gaining more marks. Due to the size of the town, the field survey had to be focused on a selected part of the town. I based this on the census data above, and selected three areas that would provide the best results showing segregation. The three areas were C, E and T (see Table 6).

A system of systematic sampling was used to collect the data. A series of points was drawn up in each of the wards being surveyed to give a fairly regular and even spread of assessment locations on streets in the areas. The data collected were then processed by the production of a dot map showing total values. An isoline map was drawn using these values to show the spatial pattern. Other information collected in the field was done by taking photographs [not provided].

Table 8 The environmental survey used

Feature	Penalty points	Maximum score	Feature	Penalty points	Maximum score
Landscape quality			**Noise**		
Trees and well kept grassed areas	0		Quiet	0	
Few trees and/or unkept grass spaces	4		Some noise	2	
No trees or grassed spaces	8	8	Noisy	5	5
Derelict land			**Air pollution**		
None	0		No offensive smells or obvious air pollution	0	
Small area	4		Offensive smells and obvious air pollution	10	10
Large area — a major eyesore	10	10			
Litter/vandalism			**Access to public open space**		
No litter, no vandalism	0		Access to park/public space within 5 min walk	0	
Some litter/vandalism	4		No park/public open space within 5 min walk	3	3
Very untidy; much vandalism	8	8			
Industrial premises			**Access to shops and primary school**		
All residential	0		Both within 5 min walk	0	
Some industrial premises	5		School only within 5 min walk	2	
Mainly industrial properties	10	10	Shops only within 5 min walk	3	
			No shops/primary school within 5 min walk	5	5
Traffic flow					
Normal residential traffic	0				
Above normal traffic	3				
Heavy vehicles and through traffic	6	6			

Results

The Lorenz curve of percentage NCIs against percentage area [not provided] *shows that the New Commonwealth Immigrants (NCIs) are located together with 77% of their total population living in 5% of the area of the town. It is therefore obvious to expect ethnic segregation to take place.*

The choropleth maps [not provided] *of mean gross household income and percentage of unemployed clearly show that a pattern has emerged in the town: in and around the inner city there is obvious deprivation, with low numbers for mean gross income and a high percentage of unemployed. The located pie chart map* [not provided] *also shows that the majority of NCIs live in the areas of high deprivation. They have located here because it is cheap and they will feel socially accepted by moving here. This therefore leads to ethnic segregation.*

The isoline maps [not provided] *also show a trend, in that as you move away from the centre of these areas the level of deprivation as measured by my survey reduces. The majority of NCIs live in these wards, hence showing segregation and also showing signs of serious deprivation.*

My photographs [not provided] *clearly show the deprivation in some of the wards, with large areas of run-down terraced housing and small shops. I have labelled my photographs to show features of the deprivation.*

Analysis

It is obvious that town X exhibits signs of segregation. The wards that show this the best are the wards with the highest percentage of ethnic minorities located in the area — the wards E, P and T— and these wards are located together (see Figure 5).

The statistics show that in these wards the percentage of NCIs is a lot greater than in the area of M for example. Another factor is shown by the other indicators of social deprivation. They show that the areas with the highest percentage of NCIs have low average income (below the average for the town) and also have the highest numbers of unemployment (almost double that of the town).

Analysis from the isoline map must be made with caution as there are obvious limitations and problems. The scoring of the points is down to personal opinion, and the interpolation of the lines is also down to the individual and so human error could take place.

Overall evaluation

The evidence suggests that there is ethnic segregation in town X. This is what I first predicted and it has been shown by my results. Segregation is when one community is apart from another community, as the Asian community is in town X. As the urban population has grown, so the segregated area has expanded. The Asian communities of town X have enlarged and amalgamated by subsequent population increases. While this does not 'cause' segregation, demographic change is evidently an important factor to bear in mind.

It may be that in Britain the most commonly used scale of measuring segregation, by wards, offers too coarse a spatial net to pick up the detailed locational pattern of what is a relatively small urban minority. Studies show that the West Indian population has integrated more and therefore there is less segregation. However, segregation of the Asian population is still common, especially in town X.

The concentration of Asians tends to occur in the least desirable locations within the urban environment, the least popular council housing or rented accommodation. There is no doubt that such minorities receive a disproportionate share of the very worst housing stock across the country. This is also evident in town X with the majority of the ethnic population located together in the most deprived areas such as the housing in ward T.

There is a high degree of ethnic segregation in town X. My study however does not explain why it exists.

Final comments
You might want to review this piece of fieldwork. You could ask yourself how it could have been improved further.

Enquiry 3: Does the village of Y match the characteristics of the rural area in which it lies?

Extracts from the write-up of a fieldwork enquiry based on a suburbanised village are given. The sections in italics are the words used by the student, except that some area names have been removed.

The context comes from the A2 course, and both AS and A2 skills are used.

Hypotheses
1   *The population of village Y mirrors that of Aylesbury Vale.*
2   *The employment pattern in the village matches that of Aylesbury Vale as a whole.*

Location
A detailed location was given of the study area (a village in Buckinghamshire, near Aylesbury), in terms of both maps and written description.

Characteristics of the area
Village Y has a number of services: a primary school, a church, a village combined grocery, newsagents and post office, two hairdressers, four public houses and a village hall. It lies close to a number of towns: Aylesbury, Wendover and Tring. The motorways M40 and M25 are not too far away.

The student then gave a detailed summary of the history of the area — taken from a historical publication of the village.

Relevant geographical theory
The area demonstrates a number of aspects of recent changes in rural areas. There has been depopulation of some parts of the community, as well more recent counter-urbanisation into the area. The number of farms has decreased, with an increase in the

amount of commuting from the village, more teleworking from home and more retired people moving into the village. These changes have been brought about by improvements in transport and communications, such as the internet, and by increased living standards. For many people, whether retired or commuters or self-employed people, the countryside is safer and more peaceful and offers better quality of life than the town.

Many villages have grown at an astonishing rate and have therefore lost their original character. These villages have been described as dormitory, commuter or suburbanised villages.

A table showing the general characteristics of a suburbanised village was then given — the table was taken from a textbook.

Methodology

For both hypotheses, I used census data, via the internet, and a questionnaire. I decided to hand out a questionnaire to the villagers. The questionnaire would have questions on the village, how long they have lived in the village and general things like that. I typed up a questionnaire up with some questions that I thought were appropriate. I then handed this pilot questionnaire out to people in my teaching group to see what the response was like. When I got the questionnaire back I realised that I needed to add some more questions to it. After I had added these I posted 60 questionnaires out to villagers. Later I went to collect the questionnaires back from each house.

I then examined the 2001 census and compared it with my findings. I drew some population pyramids, did a chi-squared calculation and also drew bar charts to show the differences in the nature of employment between the area and the village.

A questionnaire was included. This asked the following questions:
- (a) Please put a tick in your age group (A range of options in 10-year blocks was given)
- (b) Where do you live in the village?
- (c) What is your occupation? (A range of options was given)
- (d) How long have you lived in the village? (A range of options was given)
- (e) How many people live in your house?
- (f) Where were you born?

A brief explanation was given for each of these questions.

Results

A table of results was constructed (not provided), consisting of the answers to the questions above for each respondent. (The results for questions a, c and d are given in Tables 9, 10 and 12).

Bar graphs (not provided) were drawn of the outcomes of each of the questions (a), (c), (d) and (e) above.

Maps (not provided) were drawn to show the locations of the people in response to questions (b) and (f).

Analysis

Table 9 Age groups of the village

Age group	Number of people	Male	Female
1–10	0	0	0
11–20	0	0	0
21–30	3	2	1
31–40	12	7	5
41–50	15	7	8
51–60	18	7	11
61–70	6	3	3
Over 70	6	2	4

A population pyramid (not provided) was drawn using this information and compared to that of Aylesbury Vale (Figure 6).

Table 10 Data from the question 'How long have you lived in the village?'

	Length in village (years)			
Age	0–10	11–20	21–30+	Total
21–30	3	0	0	3
31–40	7	5	0	12
41–50	2	6	7	15
51–60	3	9	6	18
61–70	0	4	2	6
Over 70	0	2	4	6
Total	15	26	19	60

The student then completed a chi-squared test to see if there is a relationship between the age of the residents and the number of years they have lived in the village. He first stated a null hypothesis: that there is no difference between the observed distribution above with an expected set of data. For this he assumed that the distribution would be even, as shown in Table 11.

Table 11 Expected distribution

	Length in village (years)			
Age	0–10	11–20	21–30+	Total (category R)
21–30	1	1	1	3
31–40	4	4	4	12
41–50	5	5	5	15
51–60	6	6	6	18
61–70	2	2	2	6
Over 70	2	2	2	6
Total (category K)	15	26	19	60

Using these data a set of calculations for chi-squared was undertaken. The answer was 26.3. The degrees of freedom were then calculated as follows:

Degrees of freedom = (classes in category R – 1) × (classes in category K – 1)
$$= (6 - 1) \times (3 - 1)$$
$$= 5 \times 2$$
$$= 10$$

On consulting statistical tables, it can be seen that the result, 26.3, lies between the 0.01 (1%) value of 21.67 and the 0.001 (0.1%) value of 27.88 for 10 degrees of freedom. This means that the null hypothesis can be rejected. The student wrote:

I can be over 99% sure of the difference between the ages of the people living in the village and the length of time they have lived there. In other words, the old people are long stayers and the younger have just moved in.

Table 12 gives occupation data.

Table 12 What is the occupation pattern in the village?

People aged 16 and over	In the village (%)
Employed in primary	3.3
Employed in secondary	3.3
Employed in tertiary	46.7
Employed in quaternary	25
Retired	20
Housewife	1.6
Total	**100**

The figures in Table 12 were summarised. The census figures (2001) for Aylesbury Vale were then given (Table 13).

Table 13 Census figures for Aylesbury Vale, 2001

People aged 16–74	In Aylesbury Vale (%)
Economically active: employed full time	47
Economically active: employed part time	12
Economically active: self employed	11
Economically active: unemployed	2
Economically active: student	2
Economically inactive: retired	11
Economically inactive: student	3
Economically inactive: looking after home and family	6.5
Economically inactive: sick/disabled/other	5.5
Total	**100**

Again these figures were summarised.

Evaluation (extracts)

My results show that the population pyramids of the village and of the wider area of Aylesbury Vale (Figure 6) do not match. For example, in the pyramid of the village I have no children which is incorrect. I should have asked for the ages of the people in the families of each of the respondents rather than just the ages of the respondents. This has therefore given me a false comparison through a mistake that I made. The other readings, namely that there seem to be a higher proportion of 41–60-year-olds than in the Vale as a whole is again due to the fact that I only recorded the age of the people answering the questionnaire. I should also have used the same age ranges as the census and gone up in groups of 5 years.

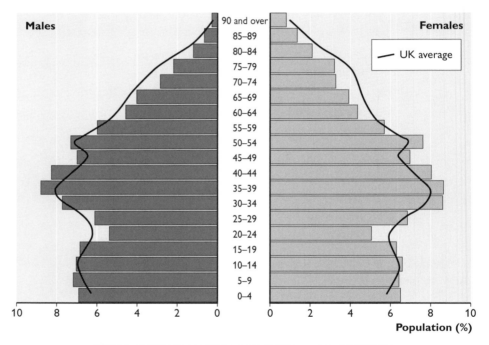

Figure 6 Aylesbury Vale population pyramid (2001)

I would perhaps have expected to see fewer children in the village, as the village is quite an old community, and more people in the older age groups as many people are retired. On the other hand I might have seen more children as the primary school is very popular with parents, though many of them live outside the village. I did however find with my chi-squared test that most of the older people had lived in the village for a long time, and that the younger residents had only moved into the village recently. To some extent this could be expected, but perhaps not if the village was a place for people to retire, in which case some of these could be recent arrivals too.

My comparison with employment patterns did not work well. I used a different classification of jobs (primary, secondary, tertiary etc.) to that of the census and consequently I could not make a good comparison. Overall therefore I have not managed to prove or disprove either of my hypotheses. If there was the chance to do this again, or if there was extra time, then I would probably change a few things.

Final comments

You might want to review this piece of fieldwork. The basic ideas and hypotheses are acceptable, and both primary and secondary data were used. However, you could ask yourself what the student did wrong, and how it could have been improved.

Example fieldwork questions and answers

A selection of responses to a range of questions is given in this section. Please note that answers have been selected to illustrate points that relate to each individual question, and each answer is identified as Candidate A, B or C. However, you should not assume that Candidate A (and likewise B and C) is the same person throughout.

Unit 2 questions

(1) The answers in this section are all in the context of rivers.

You have experienced geography fieldwork as part of the course. Use this experience to answer the following questions.

(a) Explain the geographical concept, process or theory that underpinned your fieldwork enquiry. (4 marks)

Mark scheme
Allow 1 mark per valid point in relation to theory, concepts and processes to max 4.

Candidate A
Our fieldwork enquiry was based on the concept of the hydraulic processes of rivers as we investigated the downstream variations in channel characteristics of the River X. For example, we investigated whether velocity increased as we moved further from the source. The addition of tributaries to the main channel would give the river more energy with which to erode and smooth its bed and banks by hydraulic action and attrition, making the channel more efficient so flow speed increased. This is what we would expect to find in a model river.

Candidate B
To investigate how fluvial landforms affect river channel variables across a 100–200 m stretch of the Dalby Beck in North Yorkshire. Our theory was that as cross-sectional area increased, velocity would decrease. Also the velocity would be higher on the outside bend of the river and slower on the inside bend.

Candidate C

I investigated the changes in channel characteristics on the Afon Conwy in north Wales according to Bradshaw's model. I investigated velocity, angularity of bedload, hydraulic radius and cross-sectional area. Bradshaw's model showed increases in velocity, cross-sectional area and hydraulic radius and a decrease in angularity of bedload.

> *e* Each of these answers scored 3 marks as in each case only three points are made. You need to make sure that you have covered each aspect of the task: the nature of your fieldwork and its link to concept, process or theory.

(b) How did you respond to the risks associated with undertaking your primary data collection? (4 marks)

Mark scheme
Max 1 mark for outlining the risks; max 3 marks for only one response. Responses and risks required to max 4.

Candidate A

One of the main risks was slipping on a rock while in the channel and hitting our heads so we might drown. We reduced this risk by wearing appropriate footwear with good grip and by not working in areas which were too deep or turbulent. We also worked in pairs so help could be accessed in an emergency. In addition, we minimised climatic risks such as hypothermia at the high altitudes by wearing padded coats and having first aid equipment on site if necessary.

Candidate B

There was risk of falling, slipping or drowning so we didn't enter the river if the water came over boot height and we watched where we were placing our feet. There was a risk of hypothermia so we wore lots of layers and took extra clothes and wore hats, gloves and a scarf. There was a risk of Weil's disease so we wore rubber gloves and washed our hands afterwards.

> *e* Both of these responses gained maximum credit — 4 marks

(c) Outline and justify one method of data collection that you used in your enquiry. (6 marks)

Mark scheme
Level 1 Method may be dubious in relation to the enquiry. Method is poorly explained and justification is generic or absent. 1–4 marks
Level 2 Method is appropriate and clearly outlined. Justification is specific. 5–6 marks

Candidate A

One piece of primary data we collected was velocity data. We did this using a flowmeter and we used a stop watch to time how long it took an impeller to move along its thread while in the water. A systematic sampling technique was used as readings were taken at $\frac{1}{4}$, $\frac{1}{2}$ and $\frac{3}{4}$ along the channel width at each site and each reading was done at $\frac{2}{5}$ channel depth. From the three readings an average time was calculated and this was converted into velocity (m s^{-1}) using a calibration chart. This method was used as the systematic sample gave data which were representative of the entire channel at each site and so bias or subjectivity in our technique was reduced. This gave a greater accuracy to our results so our conclusions were more reliable. The technique was also quite easy and quick to use. It gave relevant data and was safe to conduct.

Candidate B

We used systematic sampling along an interrupted line transect. We took velocity and depth readings at every $\frac{1}{10}$th across the line transect. We used this method because it is quick and easy and not biased as we didn't choose where to measure. It also shows us the expected changes across a river channel. Also it gives us a general cross-sectional area shape. We didn't take any more than 11 samples. If we did any more we wouldn't see significant changes to our results, so 11 was enough.

Candidate C

To investigate cross-sectional area we measured the width of the river along the surface from bank to bank using a tape measure. We then divided this number by 5, which is a form of systematic sampling. At these five points we measured the depth using a measuring stick with the thinnest side following the current so water did not ride up and distort the data. We calculated cross-sectional area with the equation CSA = width × average depth. We repeated this downstream. We were then able to compare our results to Bradshaw's model to justify them or not.

> ✐ The key here is to provide a 'handbook' description of the chosen method and then to state why you chose that method. Candidate A does this perfectly and gains maximum marks. Both candidates B and C lack clarity in their responses, although there is some sense of the data collection method they used. Both of these scored 4 marks.

(d) With the aid of a sketch diagram, describe one technique that you used to present data in your enquiry. (6 marks)

Mark scheme
Level 1 Simple identification/description of technique, lacking in detail.
Max Level 1 if no sketch. 1–4 marks
Level 2 Detailed description of technique, with good support from sketch. For full marks there should be some reference to actual data. 5–6 marks

Candidate A

[Good diagram of scattergraph given.]

One method used was that we converted velocity data into discharge data using discharge = cross-sectional area × velocity. We then plotted this on a scattergraph with discharge on the *y*-axis and distance from the source on the *x*-axis. Once the points for each site were plotted a smooth trend line was drawn through the points to see if there was any correlation and if it was strong or weak. Points far from this line were then circled as anomalies and omitted from our analytical techniques.

Candidate B

[Good diagram of cross-sectional section across a river given with actual figures.]

To present the data I constructed a cross-section of one of the sample points. Using scales of 2 cm to 1 m for the horizontal distance and 5 cm to 1 m for the vertical distance I plotted the points depicting the shape of the channel and joined these with a smooth curve representing the river bed. The number of squares in the cross-section were then converted and used to calculate the actual cross-sectional area of the river.

Candidate C

[Poor diagram of scattergraph given.]

We used a scattergraph to present our data. We put the independent variable on the *x*-axis (cross-sectional area) and we put the dependent variable on the *y*-axis (velocity). We then plotted our points on and drew in a line of best fit. This showed a negative correlation between velocity and cross-sectional area. It also shows anomalies clearly.

> As in question (c) the key here is to give a 'handbook' description of the technique which enables the examiner to see exactly how the task was done. Another important aspect is to make sure you refer only to presentation, and do not wander into analysis. That is a different question that could be asked. Candidate B gained full marks, and Candidate A also accessed Level 2 but failed to give any real data as required for maximum marks. Candidate C gained 4 marks because of lack of clarity and detail.

(e) Making specific reference to your results, suggest how your enquiry could be improved. (5 marks)

Mark scheme

Level 1 Improvement either absent or basic, offered only as statements without clearly explaining how improvement would come about. No obvious reference to results. 1–3 marks

Level 2 Improvement is elaborated to show how change would impact positively upon study. Thorough understanding demonstrated. Clear reference to actual results. 4–5 marks

Candidate A

Firstly, in terms of bedload size and shape we found that there was no correlation as we moved downstream with our Spearman correlation coefficient giving a value of +0.38. This may have been because our sampling technique involved randomly choosing rocks and this involved bias as people tended to choose large rocks. Therefore to improve this we could use quadrats in which a rock at a certain point must be picked up so as to make our results less subjective. Furthermore, our data in terms of our aim were limited as we only used eight sites and these were all in the upper course of the river and so we could not comment on the entire river system. To improve this we could use more sites, and larger sample sizes which cover the river equally from source to mouth to give more representative data. To make a more specific comment we could also repeat the study at different times of the year.

Candidate B

We could have used digital recording equipment for measuring velocity as it would be more accurate because there would be no human error. Also we could have gone back and repeated our data collection every month to get more reliable results as you can't say we would get the same measurements every day. There could have been heavy rainfall the day before we got there which would affect discharge levels which in turn would affect velocity. The same person should have measured one variable at each site because different people measure different ways so we would all get different results measuring the same thing.

Candidate C

The enquiry could be improved by repeating the whole experiment again to get averages of the data so statistical tests such as Spearman rank could be carried out more accurately and so that we can see if correlations are that strong with confidence tables. Using averages means that any anomalous results can be discarded. The data all needed to be collected at the same time as weather and seasons change the characteristics of the river so rather than doing one at a time and moving downstream later, have many groups all at points downstream and collect data at the same time.

> 🖉 This is a challenging question and to access Level 2 you must make sure you respond to both elements of the question: improvements and results. Only Candidate A does this and gains 4 marks. This answer could have referred more to how improvements would have impacted on results. Both Candidates B and C gained 3 marks. In neither case were actual results referred to.

(2) Here is another set of questions, with all of the answers written in the context of a human geography investigation.

You have experienced geography fieldwork as part of the course. Use this experience to answer the following questions.

(a) (i) Describe the location of your fieldwork and outline why this was a suitable site for your investigation.   (4 marks)

Mark scheme
One mark per valid point.

Candidate D

We chose Morecambe for the location of our fieldwork on social deprivation. It is a seaside town, 8 km from Lancaster town centre. Like many other seaside towns Morecambe has suffered decline due to lack of investment and cheap holidays to warm climates overseas. It has therefore become run down with many businesses having to close which has resulted in high unemployment. It was a good area to study the social deprivation because it has lots of decline and lots of areas with high unemployment. Morecambe has lots of derelict buildings and lots of areas with social deprivation. Morecambe is also a small town and therefore we could get a large spread of data given our time restraints. Transport is also good which means we could easily get around. It was also good to see the impact of regeneration schemes in Morecambe.

> *e* The candidate gives a clear 'sense of place', and outlines why the area is suitable for the investigation, albeit in a repetitive fashion. The candidate has 'set the scene' effectively, and gains full marks.

(ii) State one hypothesis or research question or issue for evaluation that you have investigated in 2 (a) (i). Describe one method of primary data collection used in this investigation. (5 marks)

Mark scheme
Level 1 A vague hypothesis/research question/issue which may have tenuous links to the specification. Methodology simplistic and/or vague, lacking sound geographical basis with obvious omissions.   1–3 marks
Level 2 A clear hypothesis/issue/research question. Methodology detailed, full and able to be replicated.   4–5 marks

Candidate D

Does the West End of Morecambe have a higher level of social deprivation than Tarrisholme? We used environmental surveys to measure the social deprivation in Morecambe. We visited two super output areas in Morecambe, one in the West End and one in Tarrisholme. In each of these areas we used systematic sampling to mark 10 points in each area along a line of transect. We visited each of these points and conducted an environmental survey in each. The environmental survey consisted of a series of questions (such as state of repair). It featured a bi-polar scoring system where qualitative data could be converted into quantitative data and therefore we could give each point in each super output area a score.

> *e* Again the candidate is very clear in the statement of hypothesis, and the means by which the data were collected. Although a little more detail could have been given regarding the nature of the environmental survey, the answer still gained maximum marks according to the mark scheme.

(iii) Discuss the limitations of your chosen method in 2(a) (ii). (6 marks)

Mark scheme
Level 1 A very simplistic awareness of the limitations, perhaps limited to one or two statements, lacking detail.   1–4 marks
Level 2 A more complex response which shows understanding of the limitations in the context of the enquiry. Detailed limitations linking to subsequent impact upon results or other aspects of the enquiry.   5–6 marks

Candidate D

The environmental survey featured lots of questions which were subjective in many parts. The sampling method that we chose could have inadvertently picked up a systematic bias in our results and therefore the houses that we surveyed might not have been representative of Morecambe. As well as this, large amounts of data can be missed which means that we didn't get a good spread of data. The environmental survey relied on qualitative data to measure the social deprivation which is hard to score and turn into quantitative data. As well as this, it was very time consuming to survey every house along every point. Lots of points marked by our systematic sampling method were not in residential areas and therefore our data might have been unrepresentative.

> As a consequence of a fairly simplistic method of data collection, as described in (a) (ii), the candidate can only offer simplistic statements of limitations here. The candidate can only criticise the technique, rather than the method, of data collection. One could ask why, if the technique had such limitations, it was chosen and used in the first place. The candidate needs to offer more sophisticated understanding to access Level 2 here, by suggesting what could be done, or was done, to improve matters. Level 1: 3 marks.

(b) Outline and justify the use of one or more techniques used to analyse your results. (5 marks)

Mark scheme
Level 1 A very basic response with more emphasis on description of a technique/s. Technique/s may not be appropriate to the results/data. Little or no justification.   1–3 marks

Level 2 Clearly focused upon outlining the technique/s. Technique/s appropriately applied to the data collected. Justification relates to the benefits/appropriateness of the technique/s in relation to the results/data or in rejection of other techniques. There must be clear justification for Level 2.   4–5 marks

Candidate D

I used a choropleth map to analyse my results. This involved marking the various scores for the social deprivation in different areas in Morecambe on to a map with different shading representing different values. This was a good method because it gave me a quick visual impression of the patterns of inequality in Morecambe and it showed me where the most and least deprived areas of Morecambe were in relation to the town centre. I also used a scatter graph to analyse my results. This used two scales to plot environmental quality scores with distance from the town centre. I then drew a line of best fit. This showed me the general trend in the relationship and was quick and easy to produce and it allowed me to use all of my data. It also enabled me to identify the anomalous results in my data so that I could try to explain them.

The candidate opts to write about two techniques as allowed by the question. In each case, the technique is described briefly and justified. This is a good response — full marks once again.

(c) Drawing upon your findings, explain how your enquiry improved your understanding of the topic. (5 marks)

Mark scheme

Level 1 Findings may be suspect with a sense that the enquiry has not been effectively understood or perhaps even carried out in full. Generic summary linked to theory in a superficial descriptive context. Simplistic statements. 1–3 marks

Level 2 Findings clearly based upon enquiry undertaken. Shows awareness of how findings link to theory. Perhaps shows understanding of the complex nature of the local environment and how this impacts upon findings. May suggest further areas of enquiry/research. 4–5 marks

Candidate D

The investigation meant that I found out more about social deprivation and inequalities in an area that has suffered decline and unemployment. My results showed me that large areas in Morecambe have been affected by the decline of tourism in the area. Large areas were run down and houses were in poor condition as people don't have the money to maintain them. My findings also showed me that some areas were not affected e.g. Tarrisholme which did not suffer from social inequality. This area had large well-kept houses and the residents were wealthy with large, neat gardens showing that social deprivation does not affect all areas.

The areas most affected were closer to the town centre of Morecambe and these have suffered most decline because the residents have become unemployed due to the closure of businesses. Areas that weren't affected were in the suburbs where the residents were not employed in the tourist industry and therefore were not affected and could maintain their house. I learnt that social deprivation does not affect all areas but it does mean that people's quality of life is affected when they are unemployed and live in poor environments. People therefore suffer worse health in deprived areas.

This answer again accessed Level 2 (4 marks) as the findings presented clearly relate to the enquiry as stated at the outset. However, there is no recognition of complexity, perhaps because the enquiry itself is fairly straightforward, and neither are any precise results offered such as the actual summative measures of social deprivation as calculated by the survey described in (a). The whole exercise is a purposeful activity and deserves its overall good mark.

Unit 4A questions

(a) Explain how the hypothesis or research question was related to the geographical theory that you studied. (5 marks)

Mark scheme

Level 1 General aim/hypothesis/RQ stated with no link to theory other than simplistic statement. 1–3 marks

Level 2 Clear aim/hypothesis/RQ stated with clear link to theory, and with strong relationship between the two. 4–5 marks

Candidate A

The aim of my fieldwork was to investigate how vegetation changes across a psammosere at North Gare, Seaton Carew. My main hypothesis was that as vegetation percentage cover increases so will the humus content of the soil. I wanted to investigate if there was a direct relationship between these two variables because they are both linked to vegetation. We have learnt how humus helps to support vegetation by providing nutrients and investigating the relationship would allow us to see if it was true for our psammosere. When studying the succession of a psammosere as they succeed inland there is an increase in vegetation cover, so we would expect there to be more vegetation and humus content further inland.

Candidate B

We studied the River Calder, a tributary of the River Lune near Lancaster. We wanted to see how closely the river conformed to the textbook model of an ideal stream. Our main hypothesis was that as distance downstream increased, the discharge of the river would also increase. We hoped to see if this was true for the River Calder as this is what is expected in the ideal stream. As you move downstream the velocity of the river should increase as the banks are eroded smoother and water volume increases and the cross-sectional area should increase due to greater levels of attrition and abrasion. All of these are important geographical processes that could be investigated with the hypothesis.

Candidate C

The aim of the fieldwork was to investigate how trampling by pedestrians affects the soil and vegetation of chalk grassland. It was carried out on a path in the Burford spur in Box Hill in Surrey. One hypothesis was that with distance from the path the number of species of vegetation would increase. This helped me to achieve my aim because trampling occurs mostly on or near the path and only few specialised species can tolerate it because it damages shoots and leads to soil compaction. However, with distance, more species can withstand the decreased rate and amount of trampling so there will be more species diversity.

Candidate D

Our aim was to understand how and why vegetation cover, species diversity and species richness change over time and distance from the sea at a psammosere succession in Morfa Harlech in north Wales.

H_0: There is no relationship between distance from sea and species diversity and richness.

H_1: As distance from the sea increases so will the species diversity and richness.

This hypothesis would help identify the different seres of the dune.

> Each of these is a good response and would have accessed Level 2. In each case the hypothesis or research question is clearly stated and the fieldwork is located well. The justification for the research is also clear. Candidates A and B would gain 5 marks, with Candidate C gaining 4. Here the 'theory' behind the fieldwork is quite simplistic. Candidate D has only given an aim, and a set of hypotheses, and has not attempted the second aspect of the question, 2 marks.

(b) With reference to a technique of data collection that you used, explain how you ensured that your collection of data was as accurate and unbiased as possible. (7 marks)

Mark scheme

Level 1 Basic identification of a technique. Simple description of the technique. Attempts to ensure accuracy and lack of bias without showing any real understanding. They may often involve the use of words without any indication that their meaning has been properly understood. 1–4 marks

Level 2 The answer describes the method clearly. The method is clearly linked to the aim of the investigation. The answer should show a clear awareness of the strengths of the strategies adopted to ensure accuracy and to avoid bias. Where relevant, it should also show awareness of the limitations of the technique and how they can be overcome or guarded against. 5–7 marks

Candidate A

To test the hypothesis a hydro-prop was used to ascertain the velocity of the river. In order for this to be accurate the same place was measured at each of the river's sites, the river's width was measured and the middle point found. The hydro-prop was placed here and measured as accurately as possible to be half way between the river's floor and surface. This was to prevent the hydro-prop's reading being distorted by the velocity being decreased near the river's floor due to erosion or at the surface due to wind or air resistance. The hydro-prop was held by one person while another independently recorded the time it took for the hydro-prop's measurer from start to finish. This was measured three times for accuracy. In order to improve accuracy it was always made sure that the hydro-prop's measurer started in the same place and was held whilst the hydro-prop was placed in the water and that the timer only started when the prop started to move. In order to make sure external factors couldn't influence the velocity reading we checked for any debris or vegetation in the river that could get caught in the hydro-prop and affect the results. We also stood in the river before beginning measuring the hydro-prop for 1 minute — this was so that any currents caused by human influence such as feet or walking were kept to a minimum and didn't affect the accuracy of the velocity measuring.

> The candidate has described the technique very clearly. The reader understands perfectly what took place and how accuracy was sought. There is also an understanding of the limitations of the technique of data collection. One may question whether enough readings were taken for the task, but despite this, the answer clearly accesses Level 2: 6 marks.

(c) Describe one technique that you used to present [*analyse for Candidate C*] your data. Explain why this was an appropriate technique to use.

(8 marks)

Mark scheme

Level 1   Basic identification of a technique. Simple description of the technique. There is no relevant reference to the particular purpose of the chosen technique. Strengths and weaknesses are not clearly understood.   1–4 marks

Level 2   The answer describes the method clearly. The method is clearly linked to the aim of the investigation. Clear reasons are given for the choice of method. Where possible it should also show awareness of the limitations of the technique and how they can be overcome or guarded against. If both strengths and weaknesses are explained well the answer will be at the top of the level.   5–8 marks

Candidate A

I used a scattergraph to present my data. I put my independent variable on my *x*-axis (% vegetation cover) and my dependent variable on my *y*-axis (humus content). I plotted my points for every site on my graph after choosing a suitable scale for both axes. I was then able to draw a line of best fit if I saw a relationship and correlation. The line of best fit goes through as many points as possible and has the same number of points above and below it. I labelled my axes and put a title on the top. This was a good technique to present my data for percentage vegetation cover and humus content because I was looking for a linear relationship and scattergraphs can present this very well as you can clearly see from the line of best fit if there is a relationship and how strong it is. The closer the majority of the points are to the line of best fit, the stronger the correlation. It was also possible to pick individual pieces of data out for specific sites and also see anomalies that had occurred if points were far away from the best-fit line. It was easy to construct and analysing it is clear and easy also. A scattergraph however does not explain the relationship and it only takes into account the two variables.

Candidate C

To analyse the data we had recorded and to see if there was any true correlation between the results we performed a Spearman rank test. I first tabulated my data with the two variables alongside each other and then ranked them with 1 being the highest rank. This simplified the data and then I calculated the difference between the ranks. Finally I squared the difference to get rid of the negative values and used the formula [*correct one given*] to produce a figure between +1 and –1. This figure was then compared to a significance table to determine the percentage it could have happened by chance. This technique was suitable because it shows the strength of the correlation between two variables. It was suitable as the figure produced allowed us to see the nature of the correlation as well because if it was a positive figure it was a positive correlation and if it was a negative figure it showed a negative correlation. The closer the figure was to +1 or –1 the stronger the correlation.

Candidate D

We started off by drawing a long profile of the psammosere at the top of the graph paper and labelled all the sites. We then drew an *x*-axis beneath the long profile, the axis being the same length as the long profile. We then drew a *y*-axis with a value of positive 50% and negative 50%. We then chose which species we would be plotting. (We took 100 pieces of data at each site so everything was already a percentage.) We then divided all our values by 2 and plotted half above the *x*-axis and half below, e.g. Site 5 = 44/2 = 22 therefore we plot 22% above the *x*-axis and 22% below the *x*-axis. Each value was plotted directly beneath its corresponding site on the long profile. N.B. If any site had a value of 0 then we plotted this point on the *x*-axis half way between that site and the previous one. Once we had plotted all the points, we then joined them up with a straight line between each site. Having completed the two lines we then shaded in the area between the two lines. This is called a kite diagram and here is an example of how it should look:

[A diagram showing the long profile of the psammosere succession and beneath it a kite diagram was given]

Kite diagrams are useful because they are a good visual aid for data but they are bad because they take a long time and can only show one species. They are also bad because they assume there is a constant rise or fall in % between the sites which may not be the case.

> 🖉 The key to any of these types of questions on techniques, whether presentation or analysis, is to give a 'handbook' description of the technique so that the examiner can clearly see how the technique was undertaken. The clearer this is the better. There is a second element to this question, requiring some justification for the technique, and this must also be addressed to access Level 2. The first two answers achieve these tasks, and both would have accessed Level 2, gaining 7 marks each. Candidate A's response could have been improved with some statement about what the outcome of the graph was. Candidate C should have taken the answer into significance levels. At A2 examiners expect students to be aware of

this for maximum credit. The response from Candidate B was similar to that of A (a scattergraph), and has not been included. Candidate D was awarded 6 marks. He clearly has an understanding of a kite diagram, and illustrated it in relation to one species. However, kite diagrams can, and often do, show a number of species, and this is an advantage as it allows clear comparison. The candidate seemed to contradict this.

(d) State the conclusions that you drew from your enquiry, making references to your results. Explain how the enquiry added to your understanding of the environment that you were studying. (12 marks)

Mark scheme

Level 1 A textbook answer with little reference to the personal study. The answer is presented in general terms with no direct reference to the candidate's own results. Any attempts to relate findings to understanding are written in the most general terms. 1–4 marks

Level 2 The answer establishes some clear connection between the results and the candidate's understanding of the environment and/or theory being studied. The answer reaches a clear and valid conclusion which is related to the aim and/or hypothesis. The candidate moves on from consideration of the hypothesis to try to explain why anomalies may not have fitted the hypothesis. 5–8 marks

Level 3 The answer is thorough and well developed. There is a clear sense of place and the candidate makes detailed reference to the actual data collected and to the conclusions that can be drawn from the data. The answer shows genuine geographical understanding of the whole fieldwork process. 9–12 marks

Candidate A

From my graph and line of best fit I was able to see a positive correlation even though it wasn't a very strong correlation. However my raw data from my investigation also show this. At site 2 my vegetation percentage cover was 31% and this increased by about two times by site 11 where it was 70%. Humus also about doubled at these sites from 0.17% at site 2 to 0.36% at site 11. This proves my positive correlation.

However site 12 was an anomaly. Site 12's vegetation percentage cover decreased by 65% from site 11's. This anomaly of 5% vegetation cover at site 12 is explained by the fact that site 12 was positioned on a footpath. This led to vegetation being trampled on by tourists and by dog walkers therefore skewing our data.

Unfortunately I was not able to statistically accept my hypothesis because my Spearman rank coefficient value was only 0.198. For 20 sites I would have needed a value of 0.38 to say it was 95% and a value of 0.85 to say it was 99% chance of there actually being a relationship. This was disappointing however, considering the factors making my psammosere a one in a million plagioclimax.

My investigation has allowed me to understand that every natural environment is not going to fit in and perfectly relate to the textbook theory. Every environment has its own individual factors that affect it giving it its own individual characteristics. For example, there is wet and dry deposition from the factory next to the psammosere which decreases the pH of the soil, killing biota. This means humus is not properly integrated into the soil so vegetation can't get nutrients. There is also an area of riprap in front of the dunes to halt the erosion. This is preventing salt spray from the sea reaching the dunes, making conditions less saline. There is also an offshore breakwater made from iron and steel slag. When the waves erode the breakwater, the iron and steel slag in the water cause a lower pH of the sea water. Due to the lower pH, vegetation such as marram grass is found much earlier. Marram grass is halophytic and so is usually found on the mobile dunes. However, it is found on the embryo dunes at North Gare. Also there is a golf course at the back of the psammosere and buckthorn planted at the end of the dunes to stop them succeeding to the golf course. This halting of the psammosere to preserve the golf course makes the length of the psammosere 250 m which is only $\frac{1}{10}$th of a textbook psammosere. All these individual factors lead to it being a plagioclimax and an individual psammosere.

Candidate B

Overall our investigation did show a general trend of an increase in discharge downstream on the River Calder. However looking at the line graph used to present our data we saw that, while the discharge did have a tenfold increase between Site 1 (0.02 cumecs) and Site 10 (0.2 cumecs), our results did show some anomalies. These were the fact that discharge peaked at Site 7 (0.46 cumecs) after which it fell to 0.39 at Sites 8 and 9, then again to 0.2 at Site 10 when we expected it to continue to rise. Site 4 also showed an odd drop in discharge. So after further investigation into the geography of the area using the Lune Environment Agency geology and water abstraction maps we managed to find what may have caused these anomalies. While the underlying geology of the whole area remained millstone grit we found that the overlying drift changed from boulder clay (a non-aquifer) to alluvium (a minor aquifer) after Site 7. This resulted in water being leached out of the channel after Site 7, reducing both the velocity and discharge of the river. Also between Sites 7 and 8 water was abstracted from the river to feed the Glasson branch of the Lancaster Canal, further reducing the water volume in the River Calder. Together these facts helped explain why the discharge fell after Site 7 instead of continuing to rise. As well as this we found that Site 10 was in fact tidal due to its confluence with the Lune estuary and this, along with its mud banks which abstracted water from the channel, explained its lower value (we must have visited at low tide). Finally just before Site 4 we found water had been abstracted for agricultural purposes resulting in the dip in Site 4's discharge (0.04 compared to Site 3's 0.12). This study has shown me how rivers will not always conform to the textbook model and can be heavily influenced not only by the geology of an area but also by humans.

Candidate C

From this investigation I found that all three of my hypotheses were correct and that all three were positive and significant correlations. The number of species increased with distance from the path. At the path only sparse plantain could be seen but the number of species more than doubled from 8 to 12 m away from the path to 16 species. After learning about soil characteristics I was able to understand more about human impacts such as soil compaction leading to bare earth and eroded footpaths in a honey-pot location. Also, the depth of the soil increased with distance from the path. It was 40 mm deep at the path but at 16 m away it was 218 mm. The correlation coefficient was 0.68 which has a significance level of 99%, showing that there was only a 1% chance the results could have occurred by chance. However, when presenting the data on a scattergraph I found that there was one clear anomaly which was where the depth was only 47 mm deep at 10 m away from the path. However I was able to apply my knowledge on soil profiles to try and explain the anomaly, for example there could have been an iron pan which impeded our measuring needle from reaching the bed rock. Finally the third hypothesis was correct so with distance from the path the plant height increases. At the path it was 51 mm but it trebled from 9 to 12 m away to 170 mm.

This investigation helped me to understand more about vegetation successions as I was able to see how the National Trust manages the chalk grassland with highland cattle and sheep, maintaining it at a plagioclimax community rather than allowing it to reach the climatic climax community of oak trees.

Candidate D

From our results we could accept our hypothesis and reject our null hypothesis. From the data we also found that % organic matter tended to increase as species diversity and richness increased. There was an exception to this though, because sites 8 and 9 (the wet slack) were below the water table and therefore were waterlogged. The sites had extremely high % organic matter, and had low species diversity due to only hydrophytes being able to survive there (creeping willow).

We found that at the embryo dune conditions were harsh with only 1.6% organic matter and only 7% vegetation cover. We also found that sand crouch grass was the dominant species here because of its halophytic nature, meaning it could survive the high salinity level.

As we moved on towards site 3 we found that the number of 'touches' for marram grass increased whereas sand couch grass decreased. This showed us that marram grass was now the dominant species and therefore marked the beginning of a new sere. Marram grass was dominant here because it is a xerophyte, meaning it is drought resistant and due to its waxy cuticle, it has decreased its water loss through transpiration making it adapted to this site.

Then as we moved to site 7 we found that the number of 'touches' of marram grass decreased from 55 to 40 and that burnet rose, a mesophyte, was now the dominant species. Site 7 had the highest species diversity and richness and had 4%

organic matter. Sites 8 and 9 however did not show the normal succession due to them being below the water line. Site 10 should have shown a climatic climax but due to the planting of a coniferous woodland there, it showed us a plagioclimax.

Through the fieldwork we did I learnt a lot about the psammosere succession and how to identify new seres as well as being able to identify when an area has been changed by human intervention. From our results we made the conclusions that as distance from sea increased so did % organic matter, and as this increased it meant that the soil was more fertile which led to an increase in species diversity and richness. We also found that the pH decreased as distance from the sea increased. Salinity of the soil also decreased as we moved further from the sea due to the lack of shells the further inland we went, meaning there was less calcium carbonate in the soil and less salt in the air.

e Each of these responses addresses the question well. In each case there are clear and explicit references to the results of the investigation and an attempt to answer the second part of the task: to state what was learnt from the experience in terms of the environment in which the work was carried out. Candidate B gained maximum marks, and Candidate A also accessed Level 3, gaining 9 marks. Candidate A's second paragraph is a little confusing, despite an excellent final paragraph. The examiner is left wondering to what extent the candidate fully understood the outcomes. Candidate C struggles with the second element of the question — how understanding of the environment was enhanced. The explanation towards the end of the first paragraph is doubtful; it is highly unlikely that a podsol would be found in such an environment — in (a) the candidate stated that the area was a chalk grassland. To some extent this answer reflects similar weaknesses to the answer in (a) where there were some concerns regarding the 'theory' behind the task. Overall Candidate C's response accessed Level 2 — 7 or 8 marks. Candidate D's answer also accessed Level 2: 8 marks. There are statements of results, and links to the theory being studied are also evident. Discussion of anomalies also features. However, the whole response lacks clarity, and is not helped by a simplistic style of writing. Depth of understanding is not apparent, and hence Level 3 could not be awarded.

Logarithmic graph paper

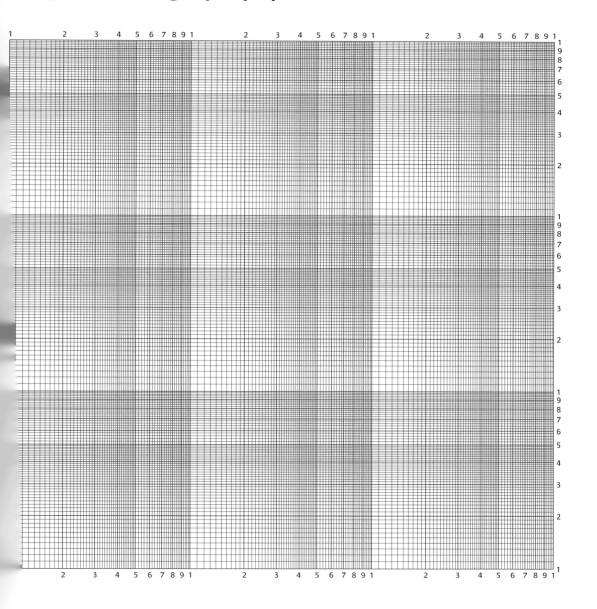

Semi-logarithmic graph paper

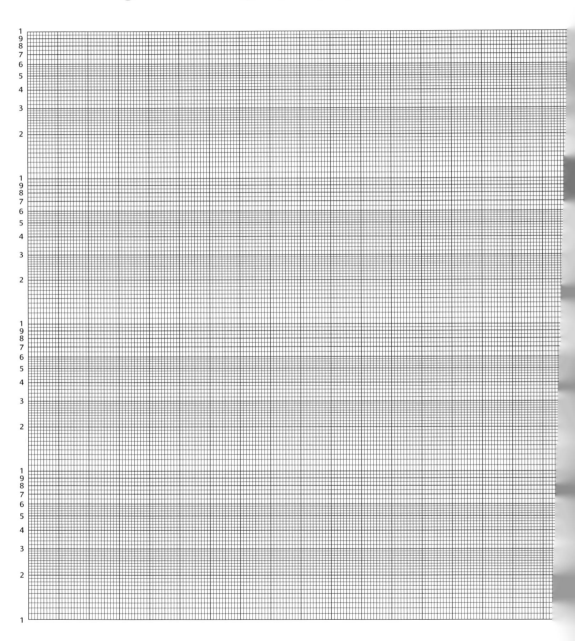

Triangular graph paper

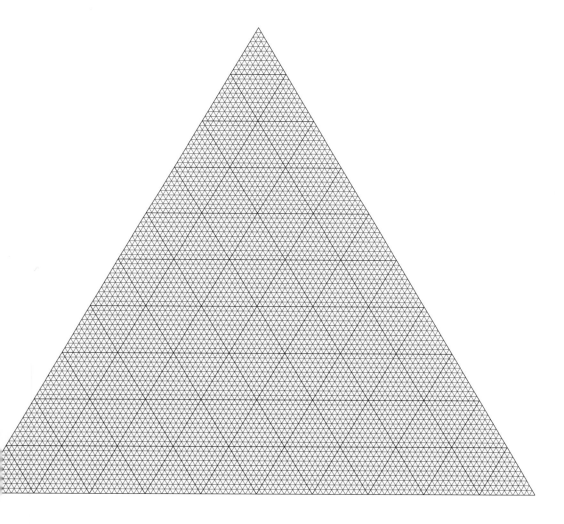